T0251727

Application of Fractals in Earth Sciences

Editor
V.P. Dimri

Taylor & Francis
Taylor & Francis Group

LONDON AND NEW YORK

ISBN 90 5410 284 5

Published by Taylor & Francis
2 Park Square, Milton Park, Abingdon, Oxon, OX14 4RN
270 Madison Ave, New York NY 10016

Transferred to Digital Printing 2007

Publisher's Note
The publisher has gone to great lengths to ensure the quality
of this reprint but points out that some imperfections in
the original may be apparent

Foreword

I am pleased to write these few lines for the book entitled 'Application of Fractals in Earth Sciences' containing 19 chapters. Since the fractal theory is new and belongs to the domain of frontline science and technology, the contents of the book are relevant and important.

The first few articles of the book pertain to the fundamentals of the fractal theory followed by its applications to earth sciences. Chapters cover application of fractals to gravity, magnetics, seismics, rock mechanics, geomorphology, climate studies and geohydrology. The eminent authors are from universities, national laboratories, academic institutions and industries their wide experience with the fractal theory and its applications should be useful to wide and diversified readership. It is hoped that this volume will fill the gap between understanding the theory and its useful applications for solving complex geological and geophysical processes.

I congratulate Dr. V.P. Dimri, Scientist of the National Geophysical Research Institute, Hyderabad for editing the book.

H.K. Gupta
Director
N.G.R.I., Hyderabad 500 007
India

Preface

The applications of fractals in earth sciences are many since many geological processes are fractal. The frequency-size distribution of earthquakes, faults, rock fragmentations, volcanic eruptions, mineral deposits etc., follows the power laws. Islands, river networks, earth topography, ocean bathymetry, erosion etc., are other examples. The sequence of stratigraphic hiatuses and energy release from earthquakes obey the fractal cantor set which had been described as the Devil's staircase. The power law empirical relations have been assumed and empirical constants are derived by fitting the model with measured values following some optimisation criteria, the simplest being least squares minimisation. These constants, called either coefficients or parameters, are denoted by a, b, c, etc. The coefficient b is normally exponent. There are many examples of b coefficients appearing in earth sciences, such as Archi's modified relation in Hydrology, called the cementation factor earlier denoted by m, or Gutenberg-Richter frequency magnitude relation in Seismology.

The 1st Chapter by Mr. Avāsthi deals with introducing fractals and its applications to problems of earth sciences. In the 2nd Chapter Prof. Rangarajan provided mathematical definition of fractals and described the self-affine fractal and a method to estimate its fractal dimension. The different types of fractals and concept of self-organised criticality have been described in 3rd Chapter by Prof. Srivastava and in the 4th Chapter he gave a detail account of percolation theory and the site-percolation threshold which allows percolation form one end of the system to another. In the 5th and 6th Chapters by Prof. Moharir, the concepts of self-similarity and multifractals are introduced, since many geological phenomena are indeed multifractal. The 7th Chapter by Dr Rita Singh present the details of processing mono and multifractal data strings which include order, information and ordered information, optical noise, fractal data strings, self-organised criticality, digital to fractal conversion, new strain of spectra etc. The fundamentals of fractal to geological problems in particular the morphology (shape) of sedimentary particles have been addressed by Prof. Sukhtankar in the 8th Chapter of this volume.

Since many geophysical data have power-law spectra, applications of fractal theory in gravity, magnetics, seismics, seismology, rock fractures and rockburst associated seismicity, climatology, topography (geomorphlogy) were dealt with remaining chapters. It was observed that density, susceptibility, reflection coefficients etc. obey fractal distribution rather than, as hitherto assumed, random distribution. Many existing methods assume random distribution of the source parameters. A new technique to estimate thickness of a sedimentary basin or basaltic lava has been proposed from gravity and aeromagnetic data in the 9th Chapter by Dr. Dimri. A scaling relation between the seismic source parameter and reflected/refracted seismic signature is discussed in the 10th Chapter by Prof. Mohan and Dr. Anand Babu. In the 11th Chapter, they observe that the phenomenon of scattering effect due to the rough surface of the earth is a beginning of the chaotic regime. The glossary of fractal words will be very useful to many readers.

Dr. Dimri introduces the concept of fractal to seismology in the 12th Chapter wherein the correlation integral method to estimate spatial and temporal correlation dimension is presented from earthquake data of the Koyna area. Dr. Srivastava gives details of the strange attractor for seismological data particularly for Koyna, the North-East and Himalayan regions in the 13th Chapter. Dr. Teotia presents the 14th Chapter wherein he has applied the concept of multifractals to seismological data. Drs. K. Shivakumar and M.V.M.S. Rao have described the role of fractals in the study of rock fractures and rockburst associated seismicity with particular reference to the Kolar gold mines, India (15th Chapter). The application of fractal dimension to flow studies has been highlighted in the 16th Chapter by Dr. Dimri.

Dr. Tiwari discusses the application of fractal and catastrophe models to climate change, earthquakes, mass extinction etc. in the 17th and 18th Chapters, presenting the basic algorithms to detect the strange attractor, second order Kolmogorav entropy, Lyapunov exponent etc. The application of fractal dimension in studying geomorphic process—a case study from historic climate data is described in the 19th Chapter by Dr. Sant.

Some readers may find repetition of definition of a few terms and mathematical expressions such as correlation function used by different authors. But readers may find that these were introduced differently citing new examples experienced by authors.

I take this opportunity to express my gratitude to all contributors of this book. I am indebted to Dr. H.K. Gupta, Director of NGRI for continued encouragement and writing foreword for this book. I thank my colleagues of NGRI, particularly Prof. Moharir, Drs. U. Raval, Rambabu, M. Ravi Prakash, Kirti Srivastava, Abhey Bansal and Prof. Ramaprasad of Osmania University for their kind guidance and help. The support of Department of Science & Technology, Government of India, particularly advices by Dr. K.R. Gupta, Director, Earth Science System in stimulating wider studies of fractals in the country, is gratefully acknowledged.

V.P. Dimri

Contents

List of Contributors

D.N. Avasthi, SPS Consultants, C-190, Sarita Vihar, New Delhi 110 044, India.

L. Anand Babu, Department of Mathematics, Osmania University, Hyderabad 500 007, India.

V.P. Dimri, National Geophysical Research Institute, Hyderabad 500 007, India.

N.L. Mohan, Centre of Exploration Geophysics, Osmania University, Hyderabad 500 007, India.

P.S. Moharir, National Geophysical Research Institute, Uppal Road, Hyderabad 500 007, India.

Govindan Rangarajan, Department of Mathematics and Centre for Theoretical Studies, Indian Institute of Science, Bangalore 560 012, India.

M.V.M.S. Rao, National Geophysical Research Institute, Hyderabad 500 007, India.

Dhananjay A. Sant, Department of Geology, Faculty of Science, M.S. University of Baroda, Vadodara 390 002, India.

K. Shivakumar, National Institute of Rock Mechanics, Kolar Gold Fields Karnataka, India.

Rita Singh, School of Computer Science, Carnegie-Mellon University 5000, Forbes Avenue, Pittsburg, PA 15213, USA.

H.N. Srivastava, Emeritus Scientist, B-1/52, Paschim Vihar, New Delhi 110 063, India.

Vipin Srivastava, School of Physics, University of Hyderabad, Hyderabad 500 046, India.

R.K. Sukhtankar, Department of Geology, Shivaji University Centre for P.G. Studies, Solapur 413 003, India.

S.S. Teotia, Department of Geophysics, Kurukshetra University, Kurukshetra 136 119, India.

R.K. Tiwari, Theoretical Geophysics Group, National Geophysical Research Institute, Hyderabad 500 007, India.

AN INTRODUCTION TO FRACTALS AND THEIR APPLICATIONS IN EARTH SCIENCE

D.N. Avasthi*

INTRODUCTION

Physical objects are generally specified as one-, two- or three-dimensional, or dimensionless, i.e., of zero dimension. This is because our sense of physical perception is limited (our mental perception has no such limit). We are familiar with co-ordinate geometry, in which the dimensions of a physical object are defined in Cartesian or rectangular co-ordinates, which we know, are at a right angle to each other. In spherical and cylindrical co-ordinates, these are defined in terms of radius vector (r), azimuthal angle or longitude (φ), a semisolid angle of a right circular cone of revolution of the radius vector about the axis passing through the origin (θ). If the axis of rotation coincides with the z-axis of the Cartesian co-ordinates, which is normally done to establish relationship between the three systems of co-ordinates, (θ) is termed co-latitude. The dimensions of the object again remain only three, e.g.

In the case of spherical co-ordinates: ($r,\ \varphi,\ \theta$)

In the case of right cylindrical co-ordinates: ($r,\ \varphi,\ z$).

All orthogonal curvilinear co-ordinate systems have only three dimensions. Physical events, on the other hand, can have more than three dimensions. For example, force has four dimensions (MLT^{-2}) and energy

* C-190, Sarita Vihar, New Delhi 110 044, India.

five dimensions (ML^2T^{-2}). The dimensions of a physical object are, for convenience, described in integral numbers. But not all objects in nature can be tied down to this simplified system of description. The jagged surfaces of mountains, the shapes of cumulus clouds, the surface of the trunks of trees with bark, the turbulence of flowing water in rivers coming down the hills, the coastal geomorphology of any landmass are some of the examples of natural objects and events which cannot be defined by integral dimensions. The co-ordinate axes to describe these objects and events may not be orthogonal and different parts of such objects and events may have to be described by different sets of co-ordinates. Such dimensions are fractional. For example, the dimensions of electrical charge are given by $M^{1/2}L^{3/2}T^1$ and of magnetic moment $M^{1/2}L^{5/2}T^1$. Mandelbrot (1967) used the word 'Fractal' to describe the characteristics of such objects and events which have fractional dimensions.

While the objects and events having integral dimensions have smooth contours, fractal dimension denotes the roughness of the object or event and can be used as a measure of the same.

A very significant property of the fractals is their scale invariance. Scale invariance appears to be a universal phenomenon in all branches of Earth Science, hence the importance of the study of fractals. Fractal distribution can be used as a means of quantifying scale invariance distribution in the phenomena studied in Earth Science. A field geologist takes the photograph of a rock feature or sequence by including a coin, a pen, a scale or his hammer to scale the feature of sequence. Now, whatever be the scale of magnification or reduction of the photograph, he can read correctly the dimensions of the feature or sequence with the help of the reference scale provided by the coin, pen, or hammer within the photograph.

Let us consider the jagged profile of the Mussoorie hills as seen from Dehra Dun, or the east or west coast of the Indian peninsula in a five million sheet map, or the silhouette of cumulus clouds and try to measure the length of the curved and jagged outline in each case. It can easily be seen that as the unit of the measuring scale is progressively decreased, the length of the outline measured increases. In other words, the larger the unit of measuring scale, the smaller the measured length and vice versa. Mathematically, we may say that the length of a fractal object is inversely proportional to the unit of measure. Or

$$N \propto 1/r,$$

where, N is the measured length using r as the measuring unit.

It will also be seen that as the magnitude of r is progressively reduced, the value of N does not increase linearly, and beyond a certain value of r, the increase in N becomes almost imperceptible. This indicates that the above inverse proportionality follows a logarithmic behaviour. Therefore, it can better be represented by the relation

$$N \propto 1/r^D,$$

or
$$N = C/r^D$$

where C is the constant of proportionality and the exponent D of the limiting value of r_{max} (further reduction in r does not bring about any perceptible change in N) is the fractal dimension of the fractal distribution given by the above equation.

The above property of scale invariance leads to another important property of fractals. If a part of a fractal in one scale is enlarged, i.e., the unit of measurement is reduced, then it very much resembles the original of which it is a part. It shows the same fractal dimension as the original. The same process may be repeated by this part and the same result is observed; and this process can go on *ad infinitum*. This indicates that:

Parts of the fractal are self-similar. Any part of it is just the replica of the whole and is similar to it. Perhaps it is one of the wisdoms that can be seen reflected in the famous *Shloka* of sage *Vedavyasa of yore;*

Om Poornamadah Poornamidam Poornat Poornamudachyute

Poornasya Poornamadaya Poornamevava Shishyate.

Mathematically speaking, while a small enough length of a straight line can appear to approximate a segment of a curve in a fractal in a given scale, in fact, it can never be possible. Or, a fractal is continuous at every point, but nowhere differentiable.

It appears that the evolution of our universe is a fractal. Howsoever much we may try to go back in time or in space, the same pattern appears to repeat *ad infinitum*. *Yogavashitha* teaches that our universe contains another universe, which contains another and so on *ad infinitum*. Also, our universe is a part of another universe in which it is contained, which is contained in another universe and so on *ad infinitum*. Events which took place in the past during the course of the evolution of the universe, have their imprint even today, except that they would appear packed together in our range of perception. Similarly, such events appear to be increasingly separated in distant future, which are again beyond the range of our perception. When events of change take place at a rate which is difficult to cope with, catastrophe occurs. Leaving out the cosmic catastrophes of the past for the study of astronomers, we see a similar pattern in the evolution of the earth and life on it, with catastrophic events occurring in the past that can be studied by the application of fractals in Event Stratigraphy. The sequence of erosion, sediment deposition, diagenesis, metamorphism and emergence of land from sea by regression, submergence of land under sea due to transgression, has been and is being repeated in different parts of the earth in different scales; thus the entire Sedimentary Geology can be seen as a fractal. Plate Tectonics is another example of a fractal event, whose magnitude and time of occurrence differ in different parts of the earth, but the process in each case is very similar.

APPLICATION IN EARTH SCIENCE

Fractal dimensions of geophysical observations have found greatest application in correlating and predicting situations from known to unknown and, therefore, have attracted the attention of exploration geophysicists. Responses of a particular subsurface geological configuration in different kinds of borehole logs can be translated as a fingerprint for similar configuration elsewhere in a geological unit on the basis of the same fractal dimensions of the borehole geophysical responses, even where other superposed events might also be influencing the recorded geophysical response.

Geophysical signals do indicate abrupt changes, which can be likened to catastrophes. The self-similarity of the jagged responses of borehole geophysical logging in different wells in a region has been qualitatively used to correlate the same geological event. Following seismic reflections from their character matching, from trace to trace, is an example of abrupt change in seismogram trace as well as use of the self-similarity property of a fractal. These have been used qualitatively to map a lithological sequence with the discerning of structural properties of the sequence, e.g. faults, buckling, thinning and thickening etc. By determining fractal dimensions of such seismic responses in a region, stratigraphic changes in the lithological sequences can be inferred. The same is true of gravity, magnetic and magnetotelluric signals from subsurface geological configurations.

The self-similar properties of geophysical signals should be considered a simple feedback process in which the same operation is carried out repeatedly, so that the output of one iteration is the input of the next, with the introduction of a dynamic parameter C on each iteration. This parameter C describes the subtle changes in the lithological suite of which the geophysical signal is a response. Such self-similar properties of the geophysical signals can be described by fractal functions, which are both continuous and scaleless, but nowhere differentiable. The functional relationship between the fractal function and the Fourier power spectrum was derived by Berry and Lewis (1980) as

$$P = f^{-(5-2D)}$$

where P is Power, f is frequency and D is fractal dimension.

The power law form of the frequency spectrum is represented as

$$A = af^{-b}$$

where A is amplitude and a and b are constants. A comparison of this equation with the equation of Berry and Lewis brings out the following relation:

$$A \cong f^{-(5/2-D)}$$

$$b \cong -(5/2 - D)$$

$$D \cong b + 5/2$$

A special case occurs when $b = -1$, or $A=af^{-1}$; which implies that the ratio of the amplitude to wavelength of all components of Fourier sinusoids is constant over all frequencies and the magnitude of that ratio depends upon the scaling factor a. In other words, the component Fourier sinusoids are self-similar in case $b = -1$. At all other values the ratio changes as a function of frequency, which is analogous to the fractal concept of self-affinity. For values of $b > -1$, the ratio increases with increasing frequency, i.e., the signal appears rougher at finer scales. For $b < -1$, the ratio decreases, causing the signal to appear smoother at smaller scales.

While Fourier transformation requires a mathematical function as input (i.e., only one frequency range value can correspond to a given value in the functional domain), no such restriction applies in fractal analysis. In the case of Fourier analysis, a continuous infinite length stationary function is assumed, while in reality, the geophysical data set is often discrete, finite and may be non-stationary. Statistical techniques are applied to minimise errors due to the assumptions made in Fourier analysis. Similar statistical techniques need to be developed for pre-processing geophysical signals before applying fractal analysis.

Thus, so long as the fractal dimension of a set of a particular property of the geophysical signal remains constant, irrespective of the variations in a property of the geophysical signal, it signifies continuity of the causative (a rock suite, a gold-bearing vein in a schist, a diamond-bearing kimberlite, a groundwater acquifer, an oil-bearing formation etc.). A change in the fractal dimension of a set of a particular property of the geophysical signal signifies discontinuity/change of the causatives. Often, the fractal dimensions of a geophysical signal come out in multimode. Multimodal fractals are helpful in giving clues to new deposits or vital geological information when these are correlated with known features of the same fractal dimensions elsewhere, as seen for the first time in the multimode. It is for these reasons, exploration geophysicists are increasingly being attracted to fractal analysis of their data.

REFERENCES

Berry MV and Lewis ZV. 1980. On the Weierstrass-Mandelbrot fractal function, *Proc. Roy. Soc. London*, **A370**: 459-484.

Mandelbrot BB. 1967. How long is the coast of Britain? Statistical self-similarity and fractional dimension, *Science*, **165**: 636-638.

FRACTALS

Govindan Rangarajan*

INTRODUCTION

The study of fractals has become an important area of research in the past two decades. Even though fractals started out as mathematical curiosities, they now find applications in many areas of science and technology (Mandelbrot, 1983; Peitgen and Richter, 1986; Barnsley, 1988; Peitgen et al., 1992). One such area is geophysics (Turocotte, 1992; Mandelbrot, 1989; Fluegman and Snow, 1989; Hsui et al., 1993; Maus and Dimri, 1994, 1995, 1996; Rangarajan and Sant, 1997; Dimri, 1998, 1999).

Fractals typically have the following properties: (a) they have fine structure, i.e. detail on arbitrarily small scales and (b) they have some sort of self-similarity, i.e., small parts of a fractal are similar to larger parts of the fractal, which in turn is similar to the whole object. Mathematically, a fractal is defined as an object whose 'fractal dimension' (which will be defined later) is greater than its usual topological dimension. Here we restrict ourselves to a brief review of some of the basic characteristics of fractals. In the next section, through simple examples, we argue that the usual Euclidean concepts of length, area etc. are inadequate to describe fractals. In the third section, we define various fractal dimensions which are better measures of fractals. In the fourth section, we introduce random, self-affine fractals which are very useful in

---

* Department of Mathematics and Centre for Theoretical Studies, Indian Institute of Science, Bangalore 560 012, India.

geophysics. The R/S analysis to calculate the fractal dimension of such fractals is described. Our conclusions are given in the final section.

INADEQUACY OF EUCLIDEAN CONCEPTS

The study of fractals originated with an attempt by Benoit Mandelbrot to answer the following simple question: How long is the coastline of Britain? The answer turned out to be more complicated than one would have naively expected (Mandelbrot, 1967). If we use a low resolution satellite picture of Britain, we see only the gross features of the coastline and get a certain length for it. If, however, we use a high resolution picture, we start seeing all the small inlets and bays and this increases the measured length of the coastline. Equivalently, if we use a smaller and smaller length unit to measure the coastline, we get higher and higher values for its total length. For natural objects such as the coastline, the length does not keep on increasing indefinitely. Still, this is contrary to what we observe when we measure the perimeter of Euclidean objects such as squares, circles etc. In these cases, irrespective of the length unit used for measurement, the perimeter does not increase. It saturates at the value expected from Euclidean geometry. This gives us the hint that Euclidean concepts such as length are inadequate to characterise objects such as the coastline. We will soon see that such objects are fractals and are better characterized by their 'fractal dimension'.

Mathematical fractals bring out the inadequacy of Euclidean concepts even better. Consider the Koch curve. It is constructed as follows (see Fig. 1): Start with a straight-line segment. Divide it into three equal segments. Replace the centre segment with the top portion of an equilateral triangle. Repeat the process with each of the four remaining line segments. Continue this process to infinity.

The Koch curve is a self-similar mathematical object. We say that an object is self-similar if it retains its shape even after all the sides of the object have been increased (decreased) by the same scaling factor.

We can easily calculate the length of the Koch curve. Let the original length of the straight-line segment be unity. After the first step in the construction process for the Koch curve, we end up with 4 line segments each of length $1/3$ (Fig. 1). Hence the total length is $4/3$. After the second step, we end up with 16 line segments each of length $1/9$. Hence the total length at this stage is $16/9$, i.e. $(4/3)^2$. Generalising this, after n steps, the total length of the curve is $(4/3)^n$. To obtain the Koch curve, we have to let the above process continue to infinity. In other words, $n \to \infty$. Since

$$\lim_{n \to \infty} (4/3)^n = \infty \tag{1}$$

Fig. 1 First few steps in the construction of the Koch curve

the total length of the Koch curve is infinite. Unlike regular Euclidean objects with infinite length, the Koch curve fits within a finite area. Further, it has no tangents at any point.

As a second example, consider the Peano curve. It is constructed as follows: Again start with a straight-line segment. Remove the middle third portion and replace it with the object shown in Fig. 2. At the second step, repeat this for each of the remaining line segments. Continue this process indefinitely. We can use the procedure outlined for the Koch curve to compute the length of the Peano curve. Let the original length of the line segment be unity. At the end of first step, we get 9 line segments of length $1/3$ giving a total length of $9/3 = 3$. After the second step, we have 81 line segments of length $1/9$ giving a total length of $81/9 = 9 = 3^2$. In general, after n steps, the total length of the curve would be 3^n. Again, the Peano curve is defined in the limit $n \to \infty$. Hence, this curve also has infinite length.

The Peano curve has a paradoxical property: It completely fills the two-dimensional plane. In Euclidean geometry, we are told that curves are one-dimensional objects, whereas planes are two dimensional. The Peano curve manages to be both simultaneously! Thus, Euclidean concepts of dimensions fail when we study fractal objects.

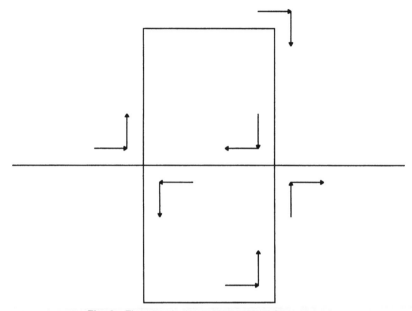

Fig. 2   First step in the construction of Peano curve

FRACTAL DIMENSION

We looked at various examples of fractal curves in the previous section. We also saw that the length of the curve is not a suitable quantitative measures for such objects. A better measure is the rate of growth of length L of the curve as a function of the size l of the measuring instrument. The mathematical relationship between the two turns out to be a power law:

$$L = A \, (1/l)^d \qquad \qquad (2)$$

where A is a constant and d is the 'measured fractal dimension'. To obtain d, we take the log on both sides of the above equation:

$$\log L = \log A + d \log (1/l) \qquad \qquad (3)$$

Therefore, if we plot $\log L$ versus $\log (1/l)$, we get a straight line whose slope gives the required value for d. One can show that the measured dimension d is related to the conventionally defined fractal dimension D by the relation:

$$D = 1 + d \qquad \qquad (4)$$

If we calculate d for the coastline of Britain using the above procedure, we get a value of 0.28. Thus, we obtain the fractal dimension to be 1.28. We notice that the coastline of Britain has a fractal dimension D which is a fractional number (i.e., not an integer) between 1 and 2. This is typical of fractals but not always true. The important point is that the fractal dimension is greater than the topological dimension (which is 1 for the coastline of Britain since it is a curve). On the other hand, if we measure D for Euclidean objects such as circles, we find that fractal dimension equals the topological dimension (both are equal to 1). A word of caution is required here. Natural objects such as coastlines etc., exhibit fractal characteristics only over a limited range of scales, i.e., the straight line behaviour in the log-log plot used to calculate d holds only for a certain range of l (measuring length).

For mathematical fractals such as the Koch curve and the Peano curve, the fractal characteristics hold over all ranges. As an example, let us calculate the fractal dimension for the Koch curve. Since the Koch curve has a reduction fraction $1/3$ at every step of the construction process, we set the size of the measuring instrument to be $1, 1/3, 1/9, \ldots$ at successive steps. In Table 1 [Bovill, 1996], we have listed the values of the measuring length l and the total length L at the first few stages in the construction of the Koch curve. Plotting $\log L$ versus $\log (1/l)$, we get a straight line with slope $d = 0.26$. This implies that the fractal dimension $D = 1.26$ for the Koch curve, which is greater than its toplogical dimension as expected.

Table 1 Data required to compute the fractal dimension for the Koch curve

Stage	l	n	L
1	1/3	4	4/3
2	1/9	16	16/9
3	1/27	64	64/27

If we repeat the same analysis for the Peano curve, we get Table 2. Plotting $\log L$ versus $\log (1/l)$, we get a straight line with slope $d = 1$. This implies that the fractal dimension $D = 2$ for the Peano curve. Note that D is an integer in this case. Thus fractals can have integer dimensions. The fractal dimension, however, is still greater than the topological dimension. This always holds for fractals.

Table 2 Data required to compute the fractal dimension for the Peano curve

Stage	l	n	L
1	1/3	9	3
2	1/9	81	9
3	1/27	729	27

The above method for calculating fractal dimension is applicable only for curves. We now briefly describe a more general method for calculating the fractal dimension. Consider an object embedded in an N-dimensional space. We cover this object with N-dimensional cubes of side l. Let the minimum number of cubes required for this be $M(l)$. Then the box-counting or capacity dimension D_c is defined as

$$D_c = \lim_{l \to 0} \frac{\ln M(l)}{\ln(1/l)} \qquad (5)$$

This implies (for small l):

$$M(l) \approx K l^{-D_c} \qquad (6)$$

Thus, D_c can be obtained by plotting $M(l)$ versus $1/l$ in a log-log plot and computing the slope of the straight-line fit (for small l).

The capacity dimension gives the usual topological dimension for Euclidean objects such as the straight line. If we cover a line segment of length L with 'cubes' of length l, the minimum number of 'cubes' required is L/l. Thus

$$M(l) = L/l = L l^{-1} \qquad (7)$$

Thus, $D_c = 1$ as expected. For fractal objects such as the coastline or Koch curve, the capacity dimension agrees with the previously defined fractal dimension. The main advantage of the capacity dimension is that it can also be used to calculate fractal dimensions of higher dimensional objects such as landscapes etc. Further, the procedure involved is quite simple (though it can be time consuming).

RANDOM, SELF-AFFINE FRACTALS

Random fractals can be generated by introducing randomness in the generating randomness process of the fractal. Many objects found in nature (montane terrain, clouds etc.) are random fractals. These objects and many natural time series records are not self-similar but self-affine. For self-affine objects, we have to scale different co-ordinates by different scale factors to obtain something similar to the original object. In particular, a self-affine function $f_H(t)$ has the following property (Voss, 1985)

$$f_H(t) \approx \lambda^{-H} f_H(\lambda t) \qquad (8)$$

where H is the so-called Hurst exponent (Hurst *et al.*, 1965). Examples of random self-affine fractals are 3-d terrain, time series records of climatic variables like temperature etc. These are self-affine (and not self-similar), because the different variables are inequivalent. For landscapes, the vertical direction plays a special role because of the presence of gravity. For time series records, the time axis is of course special.

Many random, self-affine objects found in nature best modelled mathematically by 'fractional Brownian motion' and its extension to higher dimensions. A fractional Brownian motion $f_H(t)$ is a single-valued function of one variable t (usually time). Its amplitude increments $f_H(t + \Delta t) - f_H(t)$ satisfy the following scaling property:

$$< |f_H(t + \Delta t) - f_H(t)|^2 > \approx \Delta t^{2H} \qquad (9)$$

where $< \cdots >$ is the average over many intervals of width Δt and H is the Hurst exponent ($0 < H < 1$). The special value of $H = 1/2$ gives the usual Brownian motion with $\Delta f_H^2 \approx \Delta t$.

Fractional Brownian motion shows self-affine property. If t is changed by a factor λ, then the amplitude increment Δf_H changes by a factor λ^H. Formally,

$$< \Delta f_H(t)^2 > \approx \lambda^{-2H} < \Delta f_H(\lambda t)^2 > \qquad (10)$$

Thus, it requires different scaling factors in the two co-ordinates (λ for t and λ^H for f_H). Consequently, the function is invariant under the following rescaling: Shrinking along the t-axis by a factor of $1/\lambda$, followed by rescaling of the values of the function (measured in the perpendicular direction) by a different factor λ^{-H}. For some deterministic self-affine functions, this can be done exactly. For random functions, the above considerations are valid only in a statistical sense.

For a fractional Brownian motion, the fractal dimension D is related to the Hurst exponent by:

$$D = 2 - H \qquad (11)$$

We give below the derivation of this important result, closely following the arguments given by Voss (1985). For convenience, consider a fractional Brownian motion $f_H(t)$ defined over a total time span equal to unity. Divide the t-axis into N segments of length $\Delta t = 1/N$. For each of these segments, the typical variation Δf_H in f_H is [see eqn. (9)]

$$\Delta f_H \approx \Delta t^H \qquad (12)$$

Thus, distance along the horizontal axis is Δt and distance along the vertical axis is $\Delta f_H = \Delta t^H$. Therefore, distance along the fractal curve is (the Pythagorean theorem):

$$l \approx (\Delta t^2 + \Delta f_H^2)^{1/2} \qquad (13)$$

Since there are N such segments, the total length of the curve is

$$L = Nl \approx N\Delta t \left(1 + \frac{\Delta f_H^2}{\Delta t^2} \right)^{1/2} = \left(1 + \frac{1}{\Delta t^{2-2H}} \right)^{1/2} \qquad (14)$$

On small time scales ($\Delta t \ll 1$), the second term dominates and we get

$$L \approx \left(\frac{1}{\Delta t}\right)^{1-H} \tag{15}$$

Thus, the measured dimension d is $1 - H$ [cf. Eq. (2)] and the fractal dimension D is equal to $2 - H$ [cf. Eq. (4)]. If we take $\Delta t \gg 1$, L is independent of Δt (since first term dominates in this case) and consequently

$$L \approx \left(\frac{1}{\Delta t}\right)^{0} \tag{16}$$

This implies that $d = 0$ [cf. Eq. (2)] and hence $D = 1$ [cf. Eq. (4)].

One can also define fractional Brownian motion in higher dimension. For example, consider a landscape. It is best described by a fractional Brownian motion in two dimensions (in fact, artificial landscapes in science fiction movies are often generated using fractional Brownian motion models). Let $f_H(x, y)$ denote the elevation at the point (x, y). The amplitude increments satisfy the following scaling property:

$$< |f_H(x + \Delta x, y + \Delta y) - f_H(x, y)|^2 > \approx [\Delta x^2 + \Delta y^2]^H \tag{17}$$

In this case, the relation between Hurst exponent and fractal dimension is given as:

$$D = 3 - H \tag{18}$$

The above definitions can be generalised further to higher dimensions.

To obtain the fractal dimension of objects modelled using fractional Brownian motion, it is best to first compute the Hurst exponent and then derive the fractional dimension from H (using $D = 2 - H$ in the one-dimensional case, for example). One popular way of estimating the Hurst exponent (especially in geophysics) is to use the rescaled range (R/S) analysis pioneered by Mandelbrot and Wallis [1969]. We now present the R/S analysis in its simplest form. Consider a time series of values (these could be temperatures or water levels in a reservoir or any other quantity of interest) defined over a time interval T. First, we take the whole time interval T and determine the maximum and minimum values taken by the variable (say, the temperature) over this interval. The difference between these two extreme values gives the range over this interval. Next, we divide the time interval into two equal halves and compute the ranges in each of these halves and then compute the average of these two range values. This gives one average value for the range over the time interval of length $T/2$. We then divide the interval into four equal parts and obtain the ranges in each of these intervals. Calculating the averages of these four values again gives one average value of the range over the time interval of length $T/4$. We continue this process. We end up with a set of time intervals (T, $T/2$, $T/4$, etc.) and the corresponding average values of the range for each of these intervals. Plotting this table of values in a log-log plot gives a straight line over a range of time scales if the underlying process is describable as a fractional Brownian motion. The slope of the straight-line fit gives the Hurst exponent H for the times series. The fractal dimension D is then given by $2 - H$.

One can also give a physical interpretation for the fractal dimension of the time series. If the fractal dimension of the time series lies between 1 and 1.5, it is said to exhibit 'persistence'. That is, there is positive correlation between different amplitudes in the time series. Hence the future trend is more likely to follow an established trend [Hsui *et al.*, 1993]. As the fractal dimension increases from 1.5 to 2, the process exhibits 'antipersistence'. That is, there is negative correlation between different amplitudes in the time series and a decrease in the amplitude is more likely to lead to an increase in future (and vice versa). Geophysical time records generally exhibit 'persistence'.

CONCLUSIONS

A brief introduction to fractals has been presented. The concept of fractal dimension was motivated and explained. Random self-affine fractals which play an important role in geophysics were studied using fractional Brownian motion. The R/S analysis used extensively in geophysics was described.

REFERENCES

Barnsley M. 1988. *Fractals Everywhere*. Academic Press, Boston.

Bovill C. 1996. *Fractal Geometry in Architecture and Design*. Birkhauser, Boston, pp. 195.

Dimri VP. 1998. Fractal behaviour and detectability limits of geophysical surveys. *Geophysics*, **63**: 1943-1946.

Dimri VP. 1999. Application of Fractals in Earth Sciences, editor V.P. Dimri. Oxford & IBH Publ. Co. Pvt. Ltd., New Delhi, India.

Fluegemen RH Jr. and Snow RS. 1989. Fractal analysis of long-range paleoclimatic data: Oxygen isotope record of Pacific core V28-238. *Pure Appl. Geophys.*, **131**: 307-313.

Hsui AT, Rust KA, and Klein GD. 1993. A fractal analysis of Quaternary, Cenozoic-Mesozoic, and Late Pennsylvania Sea Level Changes, *J. Geophys. Res.*, **98B**: 21963-21967.

Hurst HE, Black RP and Simaika YM. 1965. *Long-Term Storage: An Experimental Study*. Constable, London.

Mandelbrot BB. 1967. How long is the coastline of Britain? Statistical self-similarity and fractional dimension. *Science*, **156**: 636-638.

Mandelbrot BB. 1983. *The Fractal Geometry of Nature*. W.H. Freeman, New York, pp. 468.

Mandelbrot BB. 1989. Multifractal measures, especially for the geophysicist. *Pure Appl. Geophys.*, **131**: 5-42.

Mandelbrot BB and Wallis J R. 1969. Some long-run properties of geophysical records. *Water Resources Res.* **5**: 321-340.

Maus S and Dimri VP. 1994. Scaling properties of potential fields due to scaling sources. *Geophy. Res. Lett.*, **21**: 891-894.

Maus S and Dimri VP. 1995. Potential field power spectrum inversion for scaling geology. *J. Geophy. Res.*, **100**: 12605-12616.

Maus S and Dimri VP. 1996. Depth estimation from the scaling power spectrum of potential field? *Geophy. J. Int.*, **124**: 113-120.

Peitgen HO and Richter PH. 1986. *The Beauty of Fractals.* Springer-Verlag, Berlin.

Peitgen HO, Jurgens H and Saupe D. 1992. *Chaos and Fractals.* Springer-Verlag, New York.

Rangarajan G and Sant DA. 1997. A climate predictability index and its applications. *Geophys. Res. Lett.*, **24**: 1239-1242.

Turcotte DL. 1992. *Fractals and Chaos in Geology and Geophysics.* Cambridge University Press, New York.

Voss RF. 1985. Random fractals: Characterization and measurement. *In: Scaling Phenomena in Disordered Systems.* R. Pynn and A. Skjeltorp (eds.). Plenum, New York, pp. 1-11.

THE FRACTAL REDUX

Vipin Srivastava*

INTRODUCTION

Even without knowing about fractals, regardless of the subject one is interested in, if one's interest lies in knowing the nitty-gritties of nature he confronts fractals. Whether one looks at nature at the macroscopic level or inside human physiology, or towards matter at the microscopic atomic level, or even at human psychology, fractals seem to be everywhere.

Galaxies, tortuous coastlines, landscapes (mountains, rivers and forests, the mountain surface etc.), meandering rivers, snowflakes, rock formation, flow of oil and molecules in rocks etc. are some of the examples of fractals in the exterior nature around us. If we look inside the human physiology we find that the airways of the lung, the blood vessels and the neurons have a fractal character. Interestingly, aging is understood as the result of the diminishing fractal nature of these physiological components. Chaos and non-linear dynamics of fluids are inevitably connected with fractals. More recently, the microscopic world of atoms and electrons has been found to have examples of fractals in the phenomenon of percolation, quasi-crystals, Penrose tiles and even subtle objects such as electron wave functions and electronic state distributions in disordered systems. I even believe the human psyche possesses a scale-invariant fractal nature.

* School of Physics, University of Hyderabad, Hyderabad 500 046, India.

I will first introduce the meaning of fractals, their types and certain related concepts, some common examples and the definition of the fractional dimension to measure the fractality of an object. I shall also introduce the concept of self-organised criticality, a subject of growing popularity in physics, before venturing into an example wherein this could possibly be useful in connection with the earth sciences.

WHAT IS A FRACTAL

Conventionally symmetries and order in a system (such as in a crystal) have been treated as synonymous. That is, symmetries characterise an ordered crystalline system. However, there is an unconventional symmetry

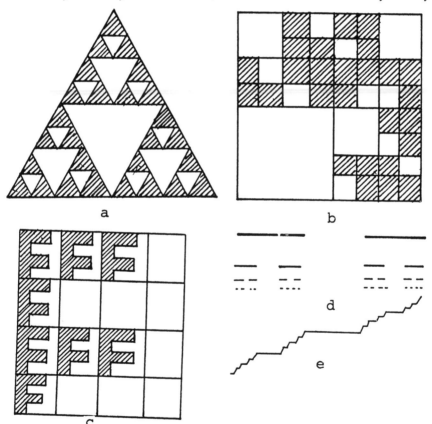

Fig. 1 Some common examples of fractals: (a) Sierpinski gasket, (b) Sierpinski carpet, (c) Curdling clusters, (d) Cantor dust and (e) Devil's staircase. One can go down in scale indefinitely to discover more and more structures of the same kind

which is possessed by certain objects ordinarily believed to be disordered—the dilatory symmetry. Parts of such an object when dilated look similar to the object itself. If this property of self-similarity under the operation of dilation exists over a sufficiently large range of length scales then the object is called fractal. Clearly, fractals have structure within structure at various length scales which are unravelled by the operation of dilation performed at various stages. Some examples of fractals are shown in Fig. 1.

Measuring self-similarity

The self-similarity possessed by a fractal object can be measured or gauged in terms of a fractal dimension which quantifies its tortuosity or the extent to which it fills the space.

Let us consider, as an example, a rugged coastline. Seen from a distance in space it will look like a wrinkled line. As the observer comes closer to earth more and more details hidden in the wrinkles (i.e., wrinkles within wrinkles) become noticeable. This amounts to viewing the coastline at continuously reducing length scale. Alternatively, if we stay on ground and try to measure the length of the coastline with the help of a yardstick ϵ, we will find something curious if we repeat the measurement with smaller and smaller ϵ to capture more and more wiggles in the coastline—we shall find that the measured length increases as ϵ decreases and indeed eventually diverges in the limit of $\epsilon = 0$. The so-called Richardson's hypothesis (Mandelbrot, 1983) states

$$L(\epsilon) \sim \epsilon^{-D} \epsilon = \epsilon^{1-D} \tag{1}$$

where $L(\epsilon)$ is the length measured with the yardstick ϵ; ϵ^{-D} is the number of steps of size ϵ; and D is some exponent.

We shall note that D is indeed a dimension. To understand this, view the coastline as a Koch curve, shown in Fig. 2(a) and look at only one side. Suppose its length is $L(\epsilon)$ on scale ϵ. If $\epsilon \to \epsilon/3$, $L(\epsilon/3)$ becomes $4L(\epsilon)/3$, i.e., $(\epsilon/3)^{1-D} = (4/3)(\epsilon)^{1-D}$, or $D = \ln 4/\ln 3$. Note that the exponent D is given in terms of 4 new length units and the factor 3 by which the scale is reduced. This form of D shows that it is a dimension—see below.

Figure 2(b) shows a unit square of side L. If the length scale is reduced by a factor of 2, we get 4 new unit squares, i.e. 2^2, where the exponent 2 is the Euclidean or the embedding dimension. If the scheme involves cutting off a new unit when the scale is reduced, then reduction of the scale by a factor of 2 will yield 3 new unit squares. That is

$$3 \text{ units} = 2^D \tag{2}$$

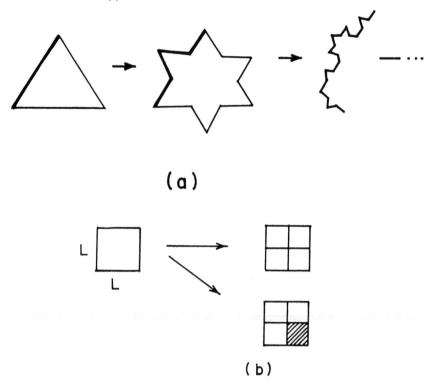

(a)

(b)

Fig. 2 (a) A coastline viewed as a Koch curve
(b) A unit square of side L yielding four new units on reduction of L by a factor of two, or three units if a new one is cut off

where D is a dimension other than the Euclidean dimension. Relation (2) leads to $D = \ln 3/\ln 2$ in terms of 3 new units produced by length scale reduction by the factor of 2.

Thus D of eqn (1) is a dimension, called fractal, or similarity, or Housdorff dimension. It is not the topological dimension. In general, if the length scale reduction by a factor of n produces m new identical units, then D is defined as

$$D = \ln m \: / \: \ln n \qquad (3)$$

A coastline would ordinarily be taken as a line, i.e., a one-dimensional entity. But we see that this is not true! It has a dimension greater than 1 but less than 2. We find that a fractal line spills over into a two-dimensional space. A fractal surface similarly spills over into the three-dimensional space by possessing a fractal dimension D which is $2<D<3$.

Length-area-volume relations

The fractal nature of an object also manifests itself in the relation between length and area, and area and volume. As a demonstration we will look at only the length-area relationship.

Recall that the standard dimensional analysis shows that

$$\text{Length} \propto (\text{area})^{1/2} \qquad (4)$$

For example, in a circle the length $2\pi R = [4\pi \,(\text{area of circular disc}, \pi R^2)]^{1/2}$. Similarly for a square, the length $4L = 4[(\text{area}, L^2)]^{1/2}$.

However, the paradoxical dimensional findings of Hack (1957) showed for the meandering rivers that their length , and the area of their basin, obey the following relation:

$$(\text{length})^{1/D} \propto (\text{area})^{1/2}, \text{ with } D>1 \qquad (5)$$

This can be cast in a general form to represent the length-area relation for fractals of dimension 1.

Consider geometrically similar islands with a fractal coastline of dimension D. We know that the ratio $(\text{length})/(\text{area})^{1/2}$ will diverge in the limit $\epsilon \to 0$. However, if we consider a G-length, measured with a yardstick of length G, and a G-area, measured in units of G^2, then the ratio

$$(G\text{-length})^{1/D} \,/\, (G\text{-area})^{1/2} \qquad (6)$$

takes the same value for all geometrically similar islands. The numerator represents the (non-standard) fractal way of determining the linear extent of an island whereas the denominator represents the standard way of doing so. The ratio can be used to determine D of a fractal curve.

As G goes to G', the ratio (6) scales by

$$(G'/G)^{1/(D-1)} \qquad (7)$$

The fractals that obey this scaling are called 'scaling fractals' while others, like Cantor dust, trees etc., are 'non-scaling fractals'.

Variants of fractals

The fractals can be generally classified as:
(i) *ordered* or *disordered* depending on whether the portions cut off (the shaded parts in the examples of Fig.1) are arranged in a regular or irregular manner; and
(ii) *scaling* or *non-scaling* as explained above; and
(iii) *uniform* or *non-uniform* depending on whether the whole object has to be represented by one or many fractal dimensions (the latter are called multifractals); and

(iv) *surface* or *volume* (or *mass*) depending on whether one is dealing with the periphery or the bulk of a fractal object.

This list may not be exhaustive.

Simple application to sedimentary rocks

The variant (iv) above is found relevant in connection with the process of diagenesis or mineral overgrowth in sedimentary rocks. The process refers to the movement of ions and molecules as controlled by the pores in the rocks. It is found that the surface of the growth is a surface fractal whose area is given as

$$\text{area} \sim L^2(L/r)^{D-2} \tag{8}$$

obtained by modelling the surface as an equilateral triangle of sides L and then successively attaching smaller and smaller triangles of sizes $L/3, L/3^2$ and so on to the middle of every straight segment of the perimeter. After n iterations there will be $N(=A\ (L/r)^D)$ segments of length $r(=L/3^n)$; the exponent D is the fractal dimension; A is of order unity. When a pore of size L is observed with a resolution r, the number of surface features will be N and the surface area will be given by relation (8).

The bulk of the growth is modelled by a volume or mass fractal such as the Sierpinski gasket. A pore of size r contributes a porosity of order $(r/L)^3$ and the total porosity is

$$\phi = C(L/r)^{D-3} \tag{9}$$

A variety of growths with a wide range of values of D have been observed. Plausible mechanisms need to be discovered to account for different values of D.

SELF-ORGANISED CRITICALITY

I am particularly keen on introducing this versatile concept from physics introduced by Bak and Chan (1991). The self-organised criticality (SOC) may be specially relevant to the happenings in nature for it takes into account many strongly interacting degrees of freedom spanning a wide range of length and time scales. In the simplest possible terms the SOC can be defined as follows:

A *self-organised critical* system is a *driven, dissipative* system consisting of

(i) a *medium* which has
(ii) *disturbances* propagating through it, causing
(iii) a *modification* of the medium, such that eventually

(iv) the medium is in a *critical state*, and

(v) the medium is *modified no more*.

By *critical state* we mean the one with no characteristic time or length scale (i.e., self-similar scaling).

The implications of the above are that

(a) large interactive systems can *naturally* evolve towards a state which exhibits self-similar scaling; and

(b) when the system is perturbed it returns to a state of marginal stability.

Note that no tuning is necessary to produce the *criticality*—it emerges automatically—hence it is called *self-organised*. The dynamics of SOC systems is intermittent, with bursts of avalanches of activity separating periods of relative quiescence. The SOC may provide a natural explanation for the ubiquitous scale-free phenomena in nature, where there is no experimentalist to tune the apparatus.

The SOC has been proven for *sand-piles* and *forest fires*. In the former sand is dropped gradually and as the height of a pile exceeds a threshold an avalanche is caused which topples the sand column. An avalanche evolves by dropping its grains on neighbouring sites some of which are 'dynamical' (which are approaching the threshold height) and others are 'absorbing'(which is analogous to 'falling off the edge'). An avalanche terminates when all sand-piles are below the threshold height after the last grain of sand in the avalanche has been dropped.

In the forest fire model one looks at *burning, burnt* (to ashes) and *green* trees. Burning trees burn for one time unit (and turn into ashes) and during this time step they ignite all neighbouring green trees. There is a non-zero probability p for a new green tree to grow out of ashes. The growth happens randomly and takes one time step. The aim is to find p for a frequency f at which fires start randomly, so that after a long transient period the trees are found to form clusters (forests) of all sizes.

The SOC should be relevant to *landscape formation* where erosion by wind, water and ice etc., plays the central role. Earthquake is another phenomenon where it should prove helpful as discussed below.

Earthquakes

The number of earthquakes, $N(E)$, for which an energy in excess of E is released has been found to obey a power law

$$N(E) \sim E^{-b} \tag{10}$$

known as the Gutenberg-Richter law. This is obeyed by the whole range of earthquakes indicating that the physics of large earthquakes must be the same as that of small and intermediate ones.

(If $E \rightarrow aE$, $N(E) \rightarrow N(aE) \sim a^{-b} E^{-b}$, i.e., the shape of the distribution remains the same showing how power laws describe scale-invariant phenomena.)

Now consider the following sequence of events in a 2-d lattice of interacting blocks in which at each block site (i,j) a force, F_{ij}, acts in the general direction of motion in a fault region: The tectonic plates slowly squeeze into each other. The forces F_{ij} build up. Eventually, failure may occur at a site (i,j) when F_{ij} exceeds a threshold value F_c, for rupture. In this process $F_{ij} \rightarrow 0$, whereas

$$F_{nn} \rightarrow F_{nn} + \alpha F_{ij} \tag{11}$$

on nearest neighbour blocks. The transfer of force to neighbours causes them to become unstable and topple, triggering their neighbours, and so on—a chain reaction or earthquake takes place (energy of the earthquake α number of distinct topplings). The earthquake stops but only for a while until the force at some other location exceeds its critical value and a new earthquake is initiated at another location. One finds that with time the subsequent earthquakes become bigger and bigger. After a long time the system self-organises into a stationary state where the distribution of earthquakes does not change with time.

Thus, the Gutenberg-Richter law can be understood as an indication that the earth has self-organised itself to a crtitical state. Now the earthquakes occur due to perturbations on the critical state. The sensitivity of the critical state to the minor perturbation ought to be understood.

REFERENCES

Bak P and Chan K. 1991. Self-organized criticality. *Scientific American*, p. 46.

Hack JT. 1957. Studies of longitudinal stream profiles in Virgina and Maryland. U.S. Geological Survey Professional Papers, 298B.

Mandelbrot BB. 1983. The Fractal Geometry of Nature. Freeman & Co., New York, p. 29.

THE PERCOLATING FRACTALS

Vipin Srivastava*

INTRODUCTION

Percolation is among the simplest of concepts in physics, yet one of the most versatile (Stauffer, 1985). Whether it is the puzzles of random walks on a lattice, or subtleties of phase transitions, or complexities of spreading electron wave function in a disordered solid, percolation comes handy. The term percolation might have been adapted in physics from the flow of fluid in porous rocks in the first place. However, having been nurtured in physics, it still seeks to address the details of the problem of wetting of rocks.

WHAT IS PERCOLATION

Percolation is about formation of an infinite connected network allowing movement from one end of the system to another, e.g., wetting of a porous rock.

Consider a discrete lattice in which the lattice sites are occupied with probability p and empty with probability $q = 1 - p$. If the two nearest neighbour sites are occupied, then a bond is formed between them (in the simple classical percolation, bonds bigger than the nearest-neighbour ones are not considered). When p is small there are isolated small clusters connected by the nearest-neighbour (nn) bonds. If the fraction p of occupied

* School of Physics, University of Hyderabad, Hyderabad 500 046, India.

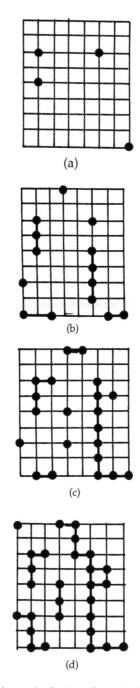

Fig. 1 Clusters growing in size as the fraction of occupied sites increases from (a) to (d)

sites—say by atoms—increases by addition of more atoms which go and occupy the sites randomly, the clusters grow in size (see Fig. 1). If this process continues, at a critical fraction $p = p_c$, an infinite network of occupied sites connected by nn bonds is formed, which allows 'percolation' from one end of the system to another. This is 'site percolation' and p_c is called the site-percolation threshold. The rules of the game can be changed slightly to consider 'nn' bonds being thrown in randomly instead of atoms to occupy the sites. In this case, we consider the probability of an nn bond being occupied or empty. The bond-percolation problem has a slightly different value for p_c.

It is important to note that the percolating network is formed only at a certain *critical* concentration of sites or bonds—p_c^S, p_c^B—which are sharply defined. The fact that these represent sharply defined critical points indicates that they mark a phase transition (Domb and Green, 1986) i.e., a qualitative change in the system occurs as the continuously varying p crosses $p_c (= p_c^S$ or p_c^B as the case may be).

Some important inequalities to note are

$$p_c^B \leq p_c^S, \tag{1a}$$

$$P^B(p) \geq P^S(p), \tag{1b}$$

where P represents the percolation probability, i.e., the probability that a site or a bond lies on the infinite network;

$$P^{tr}(p) \geq P^{sq}(p) \geq P^{hc}(p) \tag{1c}$$

where tr, sq, and hc respectively, represent two-dimensional triangular, square and honey-comb lattices; and

$$P^{fcc}(p) \geq P^{bcc}(p) \geq P^{sc}(p) \geq P^d(p) \tag{1d}$$

where fcc, bcc, sc, and d respectively, represent the three-dimensional face-centred-cubic, body-centred-cubic, simple-cubic and diamond lattices.

The following power laws represent the critical behaviours of certain important properties across the p_c :

$$P(p) \sim (p - p_c)^\beta \tag{2a}$$

$$\xi(p) \sim (p - p_c)^{-\nu}, \tag{2b}$$

where ξ is the length of connectedness ;

$$S(p) \sim (p_c - p)^{-\gamma}, \tag{2c}$$

S is the mean number of sites per cluster;

$$\sigma(p) \sim (p - p_c)^t, \tag{2d}$$

σ is the conductivity; and

$$N(p) \sim (p - p_c)^{2-\alpha} \tag{2e}$$

for concentration (per site) of clusters. The critical exponents depend on the dimensionality d of the system (i.e., they are the same for all $d=2$ lattices and the same for all $d=3$ lattices).

Ultrametric structure of percolation

It is interesting to note that percolation can be viewed as a generalisation of the problem to identify the weakest link in a chain—i.e., the crucial link that determines the strength of the entire chain. The chain is a one-dimensional structure. In higher dimensions, determination of the crucial link is actually the percolation problem. We shalll see how the clusters are a reflection of the fact that the percolation problem has an underlying ultrametric structure (Rammal *et al.*, 1986).

To understand this recall that the distance between two points, a and b, is a function $d(a, b)$ that fulfills the triangular inequality

$$d(a, b) \leq d(a, c) + d(c, b) \tag{3}$$

where c is the third point of a triangle (see Fig. 2a). An ultrametric distance measure on the other hand obeys the following inequalities *simultaneously* in a triangle *abc*:

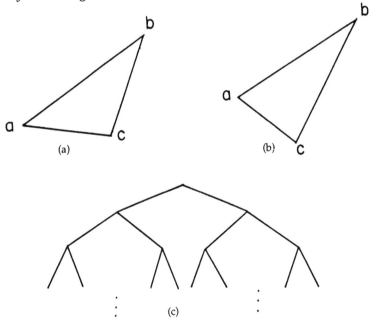

Fig. 2 (a) arbitrary triangle *abc*; (b) isosceles triangle satisfying the inequalities (4) simultaneously, (c) hierarchical nature of an ultrametric structure

$$d(a, b) \leq \max [d(a, c), d(c, b)]$$

$$d(a, c) \leq \max [d(a, b), d(c, b)] \tag{4}$$

$$d(b, c) \leq \max [d(a, b), d(a, c)]$$

That is at least two of the three distances are equal to or larger than the third (Fig. 2b). The triangle *abc* is thus either equilateral or isosceles with a short base. In an ultrametric space for a given *d*, all points having distances less than *d* can be grouped; if *d* is increased, sets join in to form bigger sets; and if *d* is decreased, sets break up into smaller ones. So an ultrametric structure has a hierarchical (or tree like) structure as shown in Fig. 2c.

In connection with percolation note that the effective percolation threshold between two points is an ultrametric measure, in that

$$p_{eff} (a, b) \leq \max [p_{eff} (a, c), p_{eff} (c, b)] \tag{5}$$

The significant implications of this are that: the clusters grow in size in a hierarchical manner; and a hierarchy of bonds governs percolation at different levels—at each level bond-sizes form a hierarchy that links clusters to merge them. Thus the global properties of a disordered system are governed by a single local property namely the bond-sizes required to link clusters of a typical size. This hierarchy of cluster sizes and their connectivity attributes self-similarity or fractal nature to the percolating cluster or its backbone. Figure 3 shows this together with the random Sierpinski carpet used to model a percolating network.

It is interesting to note that the fractal dimension *D* of the backbone of the percolating network is related to the critical exponents introduced in (2) by Stauffer and Aharony (1992) as:

$$D = d - \beta / v \tag{6}$$

where *d* is the Euclidean dimension. The cluster mass (=the number of sites in the largest cluster) is

$$M(L) \sim L^D \tag{7}$$

for a system of size *L* (i.e. typical length in one direction). More detailed studies of the percolating cluster show that it is indeed a multifractal.

APPLICATIONS TO EARTH SCIENCES

I shall mention two simple applications of the percolation idea to earth sciences—one dealing with the conductivity of the water-saturated rocks, and the other with viscous fingering, i.e., the problem of water displacing the oil in rocks.

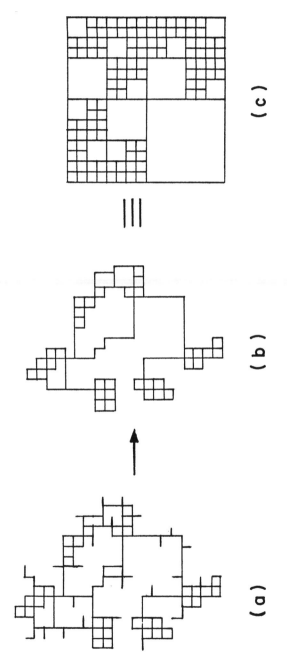

Fig. 3 (a) section of the percolating cluster; (b) backbone of the percolating cluster; (c) Sierpinski carpet to model the backbone

Conductivity in rocks

The basic empirical transport law to describe the electrical conductivity (of charge-bearing ions) σ_r in fluid-saturated sedimentary rocks is the Archie law (Korvin, 1992)

$$\sigma_r \sim a\sigma_w\phi^{-m} \tag{8}$$

where σ_w is the conductivity of water, ϕ is the porosity

$$\phi \sim (L/r)^{D-3} \tag{9}$$

L being a typical length of the rock and r, the typical pore size; the constant a is around 0.8; and the exponent m is about 2.

Relation (8) resembles the relation for conductivity in a percolating network

$$\sigma \sim (p - p_c)^t \tag{2d}$$

if

$$(p - p_c) \equiv \phi^{-1} \tag{10}$$

The exponent t is already known to be about 1.9; and p_c is practically zero for rocks since the pores are always connected, so p would represent bond fraction above the threshold value. The above analogy should indicate that the Archie law basically arises due to the pore-size distribution. Also the large value of t and its counterpart m suggest that there are large pores which reduce the conduction. This understanding leads one to believe that a 'Swiss cheese' model (shown in Fig. 4) can be constructed for the rocks.

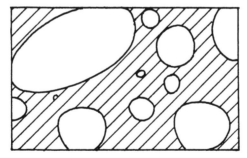

Fig. 4 The Swiss cheese model of sedimentary rocks. The voids represent the sand grains and the shade region has the pore water

Viscous Fingering

How oil and water displace each other in a rock is a question of great practical applicability. It helps in extraction of oil stored in the rocks at the bottom of the sea. It is also a problem of theoretical interest, for the water pushes the oil in a non-trivial manner.

It is well known that the hydrocarbons generated (by decay of organic matter) at a 'source rock' permeate upwards due to their low density. However, the upward movement is blocked by certain low permeable 'cap rocks' and the hydrocarbons are trapped. The region of trap is called reservoir rock and it holds both water and oil.

We need to determine saturation ratios and permeabilities, S_o and k_o of oil, and S_w and k_w of water. ϕS_o measures the oil content in the reserve and kk_o is the possible rate of extraction—here ϕ and k represent the intrinsic properties of the rock. Detailed studies using these parameters show that when an invader fluid is injected, it displaces the defender fluid oil in such a way that highly ramified finger-like patterns, called 'viscous fingering' are formed. This fractal branching is formed by the diffusion limited aggregates (DLA) (Feder, 1988). The DLAs follow the Laplace equation

$$\nabla^2 \mu = \frac{1}{D} \frac{du}{dt} \tag{11}$$

where μ is the density field and D is the diffusion constant. The control of diffusion by the Laplace equation indicates that the pressure on each point is correlated to that at all other points—hence there is intimate connection of DLA and the formation of viscous fingering with percolation and fractal etc.

REFERENCES

Feder J. 1988. *Fractals* Plenum, NY.

Korvin G. 1992. *Fractal models in the earth sciences* Elsevier, Amsterdam.

Rammal R, Toulouse G and Virasoro M. 1986. Ultrametricity for physicists. *Rev. Mod. Phys.* **58**: 765.

Stauffer D and Aharony A. 1992. *Introduction to Percolation Theory.* 2nd ed. Taylor and Francis, London.

CONCEPTS SIMILAR TO SELF-SIMILARITY IN SCIENCE

P.S. Moharir*

INTRODUCTION: SIMON'S PRINCIPLE

Self-similarity can be systematically built into designed systems. In this context, Simon's principle is relevant. Before stating it, a definition of 'system' is offered. A system is an interacting arrangement of subsystems, wherein any subsystem is a system in its own right. Consider being a system as the property. Then, we obviously see inherent self-similarity in systems. The definition of systems above is akin to the notion of a 'China box.' A China box is a box containing China box(es). There is a Channapattana doll, which is very popular. It contains a smaller doll within it. The succession of such dolls is longer as the price of the doll goes up. We frequently talk of 'wheels within wheels,' when we fail to accept or offer a simple causal explanation at any convenient level of abstraction. It is said that big fleas have small fleas on their back, and small fleas have smaller fleas still and so on. The discomfort is the same. The problem is fractility. Simon's principle asserts that a system evolves much more efficiently when there are many levels of metastable subsystems, than when there are not. It applies to natural evolution as well as deliberate design of systems.

* National Geophysical Research Institute, Uppal Road, Hyderabad 500 007, India.

Consider a problem of pattern recognition. It is believed that after some observation, one forms a rule. Then by further observation, one observes exceptions to that rule. Then one looks for a rule that covers the exceptions. Having found it, one finds exceptions to it, a rule for those exceptions, exceptions to that rule, and so on. Isn't there a fractal here?

SIGNAL-FLOW GRAPHS

In developing structures of large order in terms of structures of smaller orders, many variations are possible. Consider matrices **A** and **B**. Then, their Kronecker product $C = A \times B$ is obtained by replacing every element a_{ij} of **A** by $a_{ij}B$. Thus the structure represented by **B** is intact and in place and is replicated within a scalar multiple as many times as there are elements in **A**. The structure in **A** is retained too, but is spread over, its manifestation rate in **C** being determined by the size of **B**. This may be clarified by an example. Let

$$A = \begin{bmatrix} a & b & c \\ d & e & f \\ g & h & i \end{bmatrix}; \quad B = [\, x \quad y \,].$$

Then

$$C = \begin{bmatrix} ax & ay & bx & by & cx & cy \\ dx & dy & ex & ey & fx & fy \\ gx & gy & hx & hy & ix & iy \end{bmatrix}$$

The structures **A** and **B** can readily be seen in **C**. They can be called outer and inner structures, respectively. Other matrix products which can be used for similar purposes are the Chinese product, Khatri-Rao product, strand product etc. When any of these products is iteratively taken, the structural properties will inherit those of the component structures, the mode of inheritance being determined by the definition of the product. The properties that are propagated may be local or global. For example, if **A** and **B** are both orthogonal matrices, so would be **C**. Thus, this property can be cultivated or built into iteratively. This is self-similarity.

The Kronecker product of matrices is a very important concept in the field of linear signal processing. But the notion can be extended to cover non-linear situations (Moharir, 1992). However, the extension of the definition of Kronecker product to non-linear situations cannot take off from here. It requires some more steps.

Firstly, matrices can be used as kernels of discrete linear transforms. For the specific purpose here the kernels would be taken to be square matrices. Specifically, let $\mathbf{A}$ be $M \times M$ and $\mathbf{B}$ be $N \times N$. Then $\mathbf{C} = \mathbf{A} \times \mathbf{B}$ would be $MN \times MN$. The transforms with $\mathbf{A}$ and $\mathbf{B}$ as kernels are denoted by $\mathbf{T_A}$ and $\mathbf{T_B}$. Then

$$\mathbf{T_A} : \mathbf{X} = \mathbf{Ax}$$

and

$$\mathbf{T_B} : \mathbf{Y} = \mathbf{Ay}$$

where $\mathbf{X}$ and $\mathbf{x}$ are $M \times 1$ and $\mathbf{Y}$ and $\mathbf{y}$ are $N \times 1$. The transforms $\mathbf{T_A}$ and $\mathbf{T_B}$ can be represented in the form of signal flow graphs.

Secondly, if $\mathbf{T_C}$ is a transform based on $\mathbf{C} = \mathbf{A} \times \mathbf{B}$ as a kernel, then it has an efficient algorithm in which algorithms for transforms $\mathbf{T_A}$ and $\mathbf{T_B}$ are building blocks. This is based on the fact that $\mathbf{A} \times \mathbf{B}$ can be decomposed as a product of sparse matrices.

Thirdly, the efficient algorithm for $\mathbf{T_C}$ can be developed directly, given algorithms for $\mathbf{T_A}$ and $\mathbf{T_B}$, without decomposing $\mathbf{A} \times \mathbf{B}$ into a product of sparse matrices. This requires the notions of block-multiplexing and strand-multiplexing. In a sequence of MN samples or nodes (numbered $0, 1, ..., MN\text{-}1$), ith block involves N nodes ($iN, iN\text{+}1, ..., iN + (N - 1)$), $i = 0, 1, ..., M - 1$ and jth strand involves M nodes ($j, j + N, ..., j + (M - 1)N$), $j = 0, 1, ..., N - 1$. Block-multiplexing and strand-multiplexing use the same signal-flow-graph for every block or strand respectively.

An algorithm for the computation of $\mathbf{T_C}$ can be written down in the form of a signal-flow-graph as N-ary strand-multiplexing of $\mathbf{T_A}$ followed by M-ary block-multiplexing of $\mathbf{T_B}$. The mathematically equivalent but structurally inequivalent algorithm for computation of $\mathbf{T_C}$ can be written down as a signal-flow-graph by performing M-ary block-multiplexing of $\mathbf{T_B}$ followed by N-ary strand-multiplexing of $\mathbf{T_A}$.

The first algorithm is called a sequentially multiplexed Kronecker product algorithm for $\mathbf{T_C}$ as the algorithm for $\mathbf{T_A}$ is multiplexed first and that for $\mathbf{T_B}$ next. The second algorithm is called reverse-multiplexed Kronecker product algorithm for $\mathbf{T_C}$ as the algorithms are multiplexed in the order $\mathbf{T_B}$ and $\mathbf{T_A}$. It may be noted, however, that irrespective of whether the sequentially multiplexed or reverse-multiplexed Kronecker product algorithm is used, the algorithms for $\mathbf{T_A}$ and $\mathbf{T_B}$ are always to be strand- and block-multiplexed respectively.

Fourthly, once the sequentially multiplexed and reverse-multiplexed Kronecker product signal-flow-graphs are obtained as above, the restriction that $\mathbf{T_A}$ and $\mathbf{T_B}$ be discrete linear transforms engendered by $M \times M$ and $N \times N$ matrices could be dropped. They can now be any non-linear M-sample and N-sample transformations, and the resultant algorithms can be taken to define their sequentially multiplexed and reverse-multiplexed Kronecker products, which will now be different. In the linear case the two algorithms were mathematically equivalent because of the commutative nature of the linear operators. It can be seen that the two sections are merely interchanged in them, but the results will differ when the sections are non-linear.

Thus, here is a case of fast computational algorithms leading to generalisations which would have been unforeseen from the definitions directly. The self-similarity is produced through strand- and block-multiplexing.

CHAOS

> And now a bubble burst, and now a world.
> —*An Essay on Man*, Alexander Pope

The problem of distinguishing between noise and chaos is difficult and is related to the theory of fractals. It is frequently believed that if the dynamics of the system is driven by stochastic processes, then the fractal dimension will be infinite, and if the system dynamics is chaotic, the fractal dimension will be finite. The first of these beliefs is wrong, though the second belief is correct. Coloured noise, and in particular power law noise, for which the power spectral density varies as $f^{-\alpha}$, can give a finite fractal dimension, as was illustrated by Osborne and Provenzale (1989). Thus, the power law spectrum does not necessarily imply chaos. Frequently, finite fractal dimension is taken to imply chaos. That too, is wrong. Chaos implies finite fractal dimension. But the converse is not true.

If the fractal dimension is D, at most $2D + 1$ ordinary differential equations would suffice to rigorously characterise the system evolution. But these differential equations would be in synthetic variables and could be non-linear.

One of the quantitative studies of chaotic systems depends on Lyapunov exponents. They are plotted against some order parameter of the system. When the largest Lyapunov exponent exceeds zero, the system is said to be chaotic. But the graph of Lyapunov exponent versus the chosen order parameter is normally a fractal. At some scale, if we identify a range of the order parameter in which the largest Lyapunov exponent exceeds zero, and study this range at a greater resolution, there would be many subranges in which the largest Lyapunov exponent falls below zero. And this can happen at many scales. Hence the above graph has been called 'a lie that tells the truth.' Add to this the fact that an order parameter has also to be measured and would be subject to measurement errors. The graph then tells us even less.

Prediction in chaotic systems should be in statistical terms. Nicolis (1990) has considered this problem at length. The Markov model is a useful devise for statistical prediction. The transition probabilities in the Markov model satisfy a Chapman-Kolmogorov equation

$$P(X_\alpha, t_{i+1}) = \sum_\beta W_{\beta\alpha}\ P(X_\beta, t_i)$$

where t_i are regularly spaced time instances increasing with the index i; X_α, X_β, etc. are states of the system,

$$W_{\beta\alpha} = W(X_\alpha \mid X_\beta) = P(X_\alpha, t_\alpha; X_\beta, t_\beta)/P(X_\beta, t_\beta)$$

are the transition probabilities and $P(.)$ are the probabilities. The CK equation can be put in the form of a matrix equation

$$\mathbf{P}(X, t_{i+1}) = \mathbf{W}\ \mathbf{P}(X, t_i)$$

where $\mathbf{P}(X)$ is a column probability vector. In principle, then

$$\mathbf{P}(X, t_n) = \mathbf{W}^n\ \mathbf{P}(X, t_0)$$

is the solution, if the Markovian property holds.

A dynamic map also implies a Perron-Frobenius equation for the probability densities, involving the evolution operator of the map. That is, if the map is

$$X_{i+1} = f(X_i, \mu),$$

where μ is a parameter vector, then

$$\Phi_{i+1} = U\ \Phi_i(X)$$

where Φ is the probability density function, defined for a particular trajectory, starting at prescribed initial conditions, and U is the evolution operator derived from f. What we would be interested in is the probabilities averaged over initial conditions. That is,

$$P_{i+1} = E\Phi_{i+1} = EU\ \Phi_i(X)$$

By iteration

$$P_n = EU^nE\ P_0$$

This will define a Markov process if

$$EU^nE = (EUE)^n$$

Satisfaction of this condition will depend on the dynamics and the partitioning of the phase space. It would be useful if the cell boundaries remain invariant under dynamics. This condition is satisfied by the system in chaos.

In the systems theory, in particular, the control theory, there is a notion of observability. Some states of the system may not be observable from some output ports. In that case, nothing can be said about those states on the basis of those observations. In general, there would be a question of how sensitive or reliable the inference about some states would be to observations at some output ports. Yet, in the theory of chaos, it is uncritically believed that the fractal dimension estimated from a single time series of the system applies to and characterises the whole system. This belief does not even stand a computational test. This belief is based on a theorem due to Takens (1981) that if the embedding dimension exceeds twice the topological dimension, faithful reconstruction is guaranteed from a single time series out of the system. The approach when the single time series is available is to reconstruct a vector time series by looking at n-tuples of samples, where n is the embedding dimension. In this connection, as in many others, what must be appreciated is that the role of the theoretical results in estimation problems is not very straightforward. None of the theoretical guarantees applies, as the empirical data are noisy (Rosenstein *et al.*, 1994). As the consecutive samples of the time series are independent, principal component analysis is performed to make them so (Rosenstein *et al.*, 1994). Alternatively, mutual information between them is minimised (Fraser and Swinney, 1986). Thus, it can be seen that in actually solving the estimation problem, statistical concepts and procedures have to be invoked, and hence the guarantees based on the assumption of noise-free time series are not applicable. When one involves statistical techniques, while dealing with problems in a field which is advertised by the *butterfly effect*, can one believe that noise would not hurt much?

CATASTROPHE THEORY

One of the topics closely related to chaos is the catastrophe theory. Results in the catastrophe theory are topological, whereas for prediction in chaotic systems one needs metrical results. In this connection, it may be a tolerable or even necessary digression to know that there are many geometries. They were systematically classified by Felix Klein (1903) in his Erlangen address. In every geometry, there is a characteristic notion of a 1:1 mapping of the space into itself that leaves intact some properties of the figures in the space. The two figures in Euclidean space are said to be congruent, if there is a 1:1 correspondence between their points, which preserves distances. A 1:1 mapping of a Euclidean space onto itself, preserving distances is called isometry. If dilations and contractions are included, together with congruence, we have a notion of similarity. It recognizes angles and ratios of lengths, but not length. If shearing transformations are included, we get affine geometry. The notion of angle is dropped then and concepts of parallel lines and ratios of lengths of parallel line segments are retained. In projective geometry, all transformations suggested by perspectives are included. Mappings arising of inversions into hyperspheres and reflections into hyperplanes define an inversive geometry. Diffeomorphic transformations define a differential geometry and a set of all continuous 1:1 mappings defines a topology. A topological geometry recognizes only the continuity properties of figures. The equivalence property is called homomorphism.

The guiding urge is to be able to classify the behaviours in a finite number of possibilities. Such classifications have always had a great appeal. For example, Thom's (1975) classification theorem states that the catastrophes in two-dimensional systems with up to four parameters (more exactly, codimension of 4) can be classified in just seven (Stewart, 1975) types. This theorem has had a great influence on many looking for simplicity. But (Woodcock and Davies, 1982; Golubitsky, 1978) add another parameter and four more catastrophes arise. With yet another parameter the number goes up by three more and beyond that the number of catastrophes is infinite. Thus the simplicity of a small number of classes is not available when the number of parameters is large. The catastrophe theory is developed adequately for systems with low dimensionality (small number of variables) and deterministic chaos is studied analytically and quantitatively mostly with a small number of parameters. Rene Thom (1974, 1977) has frequently emphasised the interpretative, hermeneutic, linguistic, non-quantitative and experimental nature of the catastrophe theory and many of these attributes are likely to be apt even for chaotic systems with a large number of dimensions (or variables) and codimensions (or control parameters). Similar views have been expressed by others (Simon, 1990). We therefore believe that deterministic chaos studies should not rely entirely on analysis but also be supplemented with experiments. If catastrophes and chaos provide a language to express the experiences in terms of it, it must be realised that speaking the language and understanding the grammar are considerably different. That is, analytical understanding of chaos may neither be necessary nor sufficient. Adequacy of the analytical and quantitative methods must be tested against the experimental discoveries of various chaotic possibilities.

CHAOS IN AXIOLOGY

Johnson (1989, 1992) has recently discussed many issues, formally in the field of politics. But they are of relevance to scientists, since they offer a good metaphor for scientific concern about choosing the *best theory*. We begin with considering spaces in which there are options to choose. These options could be regarded as candidates standing for a manifesto. For scientists these could be theories or hypotheses to be ranked. The options are offered for expressing preference to the individual voters by an *agent* in a recursive pair-wise manner. For us a voter may be a criterion on which the hypotheses are judged. The agent converts the individual voter preferences into the group preference, using a rule of aggregation agreed upon by everyone. Let n be the number of voters. Then a q-rule requires minimum of q voters to prefer x over y, for that to be taken to be the group preference. Such a subgroup of voters which can decide the group preference is called the *winning coalition*.

Consider now a toy democracy in which the agent offers three candidates x, y and z. Let there be three voters A, B and C. Let their preferences be

$$A: x > y > z,$$

$$B: y > z > x,$$

$$C: z > x > y,$$

where $>$ stands for *is preferred to*. Let the agreed aggregation rule be the majority rule. The majority M has

$$M: x > y > z > x$$

Thus, whereas the preferences of individuals are assumed to be transitive, the preference of the group is patently intransitive. This is called the Codorcet paradox or the *cyclic majority paradox*. If one can come from some x back to it, following the preference operator $>$, there is said to be a cycle, which represents intransitivity.

Arrow (1963) generalised the above example into an *impossibility theorem*. If the social choice rule is to be non-dictatorial, then intransitivity is generic. Duncan Black's (1958, 1984) median voter theorem states that if every voter votes according to the distance from his/her bliss point, the point at which the personal utility function is maximum, and his/her preference is unimodal, then for the one-dimensional choice space, the median of all bliss points dominates all other choices. The dimensionality of the choice space is the chief limitation of the Black theorem.

There are two dimensions called the Nakamura (1979) dimension D_N and the chaos or instability dimension D_C, which determine the performance of the democracy we want to establish (Schofield, 1989). Let the dimension of the policy space be D. Then each individual maps R^D into R, where R^D is a D-dimensional Euclidean space. A set of points C is called a core, if for any x in C, there is no point y in R^D which dominates it.

If $D \leq D_N - 2$, there certainly exists a core and local cycles are not possible. Then the agent would be able to take the group to some point in the core transitively, but he will not be able to take it beyond that. If $D = D_N - 1$, the local cycles can exist, but they all belong to a Pareto set. If $D \geq D_N$, the cycles may exist outside the Pareto set also. The agent may now exploit the local intransitivity. If $D \geq D_C$, cycles must exist; if any core exists it must be on the boundary of the policy space, and if the policy space has no boundary, the core is most likely to be empty. If $D \geq D_C + 1$, cycles must exist all over the policy space and the core does not exist even if the policy space has a boundary. McKelvey (1976) has said pithily that if transitivity is breached, the breach is total. The agent can then become an *agenda manipulator*. Dimensionality of the choice space can also be adjusted by the agent for his purpose.

The choice spaces of large dimension almost always lack a dominant alternative. This is a global property. Then there is a local property that in the vicinity of almost every point there are other points which dominate it. These two properties together lead to the generic instability of the majority rule.

Let q out of n constitute a winning coalition. Then (Greenberg, 1979)

$$D_N = [q/(n - q)]$$

where $[i]$ stands for the integer part of i. Thus, for a majority rule (n odd) D_N is unity. If $n = 101$ and $q = 71$, then $D_N = 2$. This would illustrate the advantage of supramajority rule.
Further (McKelvey and Schofield, 1986)

$$D_C = 2q - n + 1.$$

Thus, for majority rule $D_C = 2$. For a 55 out of 101 rule it is 10, and for a 61 out of 101 rule it is 22.

In the dynamic system the trajectory is determined by the dynamics. In the system with an agent, the path is chosen by the pair-wise alternatives provided by the agent. Existence of a core in the latter and fixed points in the former are comparable. The limit cycle in the former and a cycle in the latter are comparable too. In the theory of dynamic systems a focus becoming a limit cycle is a bifurcation. A chaotic system is required to be dense in limit cycles. In the policy spaces also ubiquity of cycles is a condition for manipulability. Thus, there is a good deal of parallelism.

Following the tradition of axiomatic development, Gibbard (1973) and Satterthwaite (1975) proved a theorem that any voting system satisfying some standard norms is also manipulable. That is, strategic behaviour is permissible in voting systems. Ease of manipulation is a more relevant consideration than openness to manipulation. This makes it a problem of complexity theory. To reduce the chances of successful manipulation, the system should be chosen to make the manipulation most difficult.

Now coming back to the problem of choosing the best hypothesis (candidate), when there are many criteria (voters) whose preferences can be computed, how can we be sure that the author we are currently reading is not working as an agent and manipulating the agenda?

ACKNOWLEDGEMENTS

I am grateful to Shri V.M. Maru for intellectual interaction, to Dr. V.P. Dimri for inviting me to write this paper and to the Director, NGRI, for material and motivational support.

REFERENCES

Aki K. 1981. A probabilistic synthesis of precursory phenomena. *Earthquake Prediction*, pp. 566-574. Simpson D.W. and Richards P.G. (eds.), Amer. Geophy. Union, Washington, D.C.

Arrow KJ. 1963. *Social Choice and Individual Values*, Yale Univ. Press, New Haven.

Black D. 1958. *The Theory of Committees and Elections*. Cambridge Univ. Press, Cambridge.

Black D. 1984. On the rationale of group decision making, *J. Polit. Econ.*, **56**: 23-34.

Frazer AM and Swinney HL. 1986. Independent coordinates for strange attractors from mutual information. *Phys. Rev.*, A33: 1134-1140.

Frisch U and Parisi G. 1985. Fully developed turbulence and intermittency. In: *Turbulence and Predictability in Geophysical Fluid Dynamics and Climate Dynamics*, Ghil M. (ed.). North Holland, Amsterdam, p. 84.

Gibbard A. 1973. Manipulation of voting schemes: a general result. *Econometrica*, **41**: 587-601.

Golubitsky M. 1978. An introduction to catastrophe theory and its applications. *SIAM Rev.* **1**: 352-387.

Greenberg J. 1979. Consistent majority rules over compact sets of alternatives. *Econometrica*, **41**: 285-297.

Hartmann WK. 1969. Terrestrial, lunar, and interplanetary rock fragmentation. *Icarus*, **10**: 201-213.

Johnson PE. 1989. Formal theories of politics: the scope of mathematical modelling in political science. *Math. & Comp. Modelling*, **12(4/5)**: 397-404.

Johnson PE. 1992. Formal theories of politics II: the research agenda. *Math. & Comp. Modelling*, **16(8/9)**: 1-13.

Klein F. 1903. Vergleichende Betrachtungen über neaere geometrische Forschungen. *Math. Annalen*, 43.

Lord EA and Wilson CB. 1986. *The Mathametical Description of Shape and Form*. Ellis Horwood Publishers, Chichester, 260 pp.

Mandelbrot BB. 1989. Multi-fractal measures, especially for geophysicists. In: *Fractals in Geophysics*. pp. 5-42. Scholz CH and Mandelbrot BB (eds.). Birkhäuser Verlag, Basel, 313 pp.

McKelvey RD. 1976. Intransitivities in multidimensional voting models and some implications for agenda control. *J. Economic Theory*, **12**: 472-482.

McKelvey RD and Schofield N. 1986. Structural instability of the core. *J. Math. Econ.*, **15**: 179-198.

Moharir PS. 1992. *Pattern-Recognition Transforms.* Research Studies Press, Taunton, pp. 38-39.

Nakamura K. 1979. The vetoers in a simple game with ordinal preferences. *Internat. J. Game Theory*, **8**: 55-61.

Nicolis C. 1990. Chaotic dynamics, Markov processes and climate predictability. *Tellus*, **42A(4)**: 401-412.

Osborne AN and Provenzale A. 1989. Finite correlation dimension for stochastic systems with power-law spectra. *Physica D*. **35**: 357-381.

Rosenstein MT. Collins JJ and De Luca CJ. 1994. Reconstruction expansion as a geometry-based framework for choosing proper delay lines. *Physica D*, **73**: 82-98.

Satterthwaite MA. 1975. Strategy-proofness and Arrow's conditions: existence and corresponding theorems for voting procedures and social welfare functions. *J. Econ. Theory*, **15**: 187-217.

Schofield N. 1989. Smooth social choice. *Math. & Comp. Modelling*, **12(4/5)**: 417-435.

Schofield N and Tovey CA. 1992. Probability and convergence for supra-majority rule with Euclidean preferences. *Math. Comp. Modelling*, **16(8/9)**: 41-58.

Simon HA. 1990. Prediction and prescription in systems modelling. *Op. Res.*, **38**: 7-14.

Stewart I. 1975. The seven catastrophes. *New Scientist*, pp. 447-454.

Takens F. 1981. Detecting strange attractors in turbulence. *Lecture Notes in Mathematics*, vol. 898. Springer-Verlag, Berlin, p. 366.

Thom R. 1974. *Models mathematiques de la morphogenese.* Ser.10/18, Union Generale D'Editions, Paris.

Thom R. 1975. *Structural Stability and Morphogenesis.* W.A. Benjamin, Reading.

Thom R. 1977. Structural stability, catastrophe theory and applied mathematics. *SIAM Rev.*, **19**: 529-533.

Turcotte DL. 1989. Fractals in geology and geophysics. In: *Fractals in Geophysics*, pp. 171-196. Scholz C.H. and Mandelbrot BB (eds.). Birkhäuser Verlag, Basel.

Woodcock A and Davies M. 1982. *Catastrophe Theory.* Penguin Books, Harmondsworth, 171 pp.

MULTIFRACTALS

P.S. Moharir*

INTRODUCTION

The way in which matter and objects occupy space needs a description (Lord and Wilson, 1986, p. 7). This has aspects of form and shape. If only the aspects of the form which relate to the external world are concerned, the word *shape* is used. The word *form* implies that some aspects of internal structure are also pertinent (Lord and Wilson, 1986, p. 8). Some descriptions are continuous, using differential and metrical properties of curves and surfaces. Others are discrete and use point-sets (e.g. in crystallography). The concepts related to fractals (Mandelbrot, 1977) have introduced a new paradigm in this field. But it is interesting to note that an average applied·scientist, who has embraced this new paradigm rather enthusiastically, is not fully aware of what it is an alternative to, or what it is a progress over. Lord and Wilson (1986) give a good treatment of the prefractal paradigms, presented in the post-fractal era, and yet untouched by fractals. Their work discusses how to describe and represent trajectories, surfaces and volumes. Earth scientists should regard these tasks as central to many of their endeavours and will be surprised to find much material in Lord and Wilson (1986) that they should have been familiar with, before coming to fractals. But let us now leap-frog to fractals.

* National Geophysical Research Institute, Uppal Road, Hyderabad 500 007, India.

DIMENSIONS OF THE PROBLEM

There is what Bellman (1961) called the *curse of dimensionality*. In classical notions, the dimensionality is one, two, three, ..., infinity. As the dimensionality increases, the generalisation of concepts becomes increasingly more difficult. Generalisation from one dimension to two is very difficult. Generalisation from two dimensions to three is also difficult, but usually less so. Thereafter it is supposed to be straightforward. This is not strictly true. Dimension is a parameter in the physics of the problem, stated in mathematical terms. Therefore, the solution can depend very sensitively on the dimension of the problem. For example, Huygen's principle holds in only one- and three-dimensional spaces and not in the two-dimensional space (Carrion, 1987). And yet all our textbook cartoons on Huygen's principle were on plane paper. This goes to illustrate that we had not negotiated even the classical concept of dimension too satisfactorily. The computational complexity of algorithms depends on the dimensionality of the problem. This dependence could be polynomial or combinatorial. It has been said (Moharir, 1992) that the *curse of dimensionality* itself is a multi-dimensional curse. And yet, to pun, there was some *integrity* to that curse in the prefractal days. Now it stands withdrawn through generalisation to non-integral fractal dimensions.

The notion of fractals adds another dimension to the curse of dimensionality. The dimension can now be fractional. A small fractal dimension is regarded as a favourable situation. That is a misconception. The computational requirement of the reliable estimation of fractal dimension itself is non-polynomial. Further, a structural study of the problem depends not only on the dimension, but also on the codimension. The upper limit on the codimension is a non-polynomial function of the embedding dimension, which itself is higher than the fractal dimension. Theoretical results about the structural classification of the problems are available only up to very small values of dimension and codimension. Therefore, small fractal dimension is not much of a source of satisfaction.

Estimating fractal dimension is a difficult problem. Geometrical fractals obtained by prescribed non-statistical rules are useful in assessing the validity of the methods of estimating the fractal dimension. One of the methods is called box counting. It is refined to a box-flex method (Barton, 1995). In these methods there is a question of the proper orientation of the boxes. Barton (1995) has shown that the fractal dimension of a straight line, which is known to be 1, is estimated anywhere between 1 and 1.06, depending on the orientation of the boxes. The chosen standard is the least favourable one, as the straight line is the worst case of geometrical anisotropy. This illustration of measurement artifacts can be juxtaposed with the measurement of the fractal dimension of the San Andreas fault

(Scholz and Aviles, 1986), wherein the estimate was 1.0008-1.0191. Can it be taken to be different from unity? The limitations of the methods for estimating fractal dimension are discussed by Brown (1995) and Pruess (1995).

The fractal dimension of the universe is estimated to be 42 (Ruellè, 1990). That is about the upper limit we may come across.

FRACTALS

A fractal is a scale-invariant structure whose statistical properties are unchanged under dilation or change of spatial length scale (Erkman and Stephanou, 1990). In this definition, the word 'statistical' is inappropriate or, at least, constraining. Therefore, it is desirable to drop it. There are many fractals, such as Cantor dust, Sierpinski carpet and Menger sponge, etc., which are non-statistically defined in terms of regular recursive procedures. The 'self-similarity' in them is actually 'self-identity'. The similarity could be defined in terms of many properties, which may even be non-numerical. The fractal is said to be a Hausdorff fractal if its Hausdorff dimension is strictly larger than its topological dimension. The topological dimension is one less than the Euclidean dimension of the space in which the structure exists. If the Hausdorff dimension does not exceed the topological dimension, but the Kolmogorov capacity does exceed the latter, the fractal is called Kolmogorov fractal (Vassilicos and Hunt, 1991). K-fractals possess local self-similarity, and H-fractals are characterised by global self-similarity.

The concepts become clear and memorable by a good turn of phrase. As an example, the *butterfly effect* (Lorenz, 1993) is quite an apt term while explaining *deterministic chaos*. A similar contribution has been made by Power and Tullis (1995) to explain the notion of *self-similarity*. They imagine *an elephant and an ant* walking on a self-similar topographic surface, and coming up with the same experience. This should straightaway suggest the relevance of fractals to the theory of evolution, which is summarised in the phrase *survival of the fittest*. Environment experienced by living species is a fractal and different species can find a conducive niche for themselves at different scales.

If the fractal dimension is close to two, the contribution to area by small particles dominates (Turcotte and Huang, 1995). Under fragmentation the volume is conserved, but the area increases. Generating this extra area needs energy. Thus, if more energy is available for fragmentation, it should suggest that the fractal dimension will increase. If the fractal dimension is close to three, the contribution to volume by the small particles is negligible (Turcotte and Huang, 1995). In the communition model for fragmentation (Sammis and Steacy, 1995) it is assumed that if two fragments of comparable size are in direct contact, one of them will break up.

ALGEBRA OF FRACTALS

The power-law relation is the only truly self-similar relation. Hence some algebra of self-similarity can be readily developed. Let

$$z = k_{zy} \, y^{-d_{zy}}$$

be a power-law or fractal relation between z and y. Let

$$y = k_{yx} \, x^{-d_{yx}}$$

be a fractal relation between y and x. Then we get

$$z = k_{zy} \, k_{yx}^{-d_{zy}} \, x^{-(d_{zy} + d_{yx})}$$

which can be written as

$$z = k_{zx} \, x^{-d_{zx}}$$

which is a fractal relationship with

$$k_{zx} = k_{zy} \, k_{yx}^{-d_{zy}}$$

and

$$d_{zx} = d_{zy} + d_{yx}.$$

Thus, in the composition of two fractal relations, the fractal dimensions get summed up, but the relation between the so-called constants is a fractal relation.

Consider again the fractal relation as follows:

Let $u = dy/dx$. Then

$$u = -k_{yx} \, d_{yx} \, x^{-d_{yx} - 1}$$

which is a fractal relation

$$u = k_{ux} \, x^{-d_{ux}}$$

with

$$k_{ux} = -k_{yx} \, d_{yx}$$

and

$$d_{ux} = d_{yx} + 1.$$

That is, fractal dimension increases by unity due to differentiation. Correspondingly, fractal dimension decreases by unity due to integration. As differentiation and integration are building blocks in linear filters, one may now be equipped to understand the passage of fractal processes through linear filters. But there is a basic difficulty in this, in that the sum of two fractal processes is not a fractal process. Hence, fractal process is not an adequate notion to deal with natural phenomena.

However, consider

$$u = k_{ux}\, x^{-d_{ux}}, \qquad v = k_{vx}\, x^{-d_{vx}}$$

be two fractal processes and let

$$y = uv$$

which is a product of two fractal processes. Then, y is a fractal process with

$$k_{yx} = k_{ux} + k_{vx} \quad \text{and} \quad d_{yx} = d_{ux} + d_{vx}$$

Hence, the dimensions add as in the case of composition of two fractal processes, but the constants merely multiply and do not combine fractally.

MULTIFRACTALS

Originally, the notion of self-similarity was defined for sets. Its limitations have been brought out above. In view of these, it was necessary to go from fractal sets to what are called multifractal measures (Mandelbrot, 1989). This is a transition from characterisation by a single number (the exponent in the power-law relation or the dimension) of geometrical objects to that by a function, which may have any number of parameters. In an equivalent statement, the characterisation is not in terms of one or two or three dimensions, but by a family of generalised dimensions D_q as a function of index q, which goes from $-\infty$ to ∞. Further, for any value of q, D_q may even be negative. The term *multifractal* was introduced by Frisch and Parisi (1985). A multifractal measure can be decomposed into the union of a continuous infinity of components. Each component is an infinitesimal fractal (unifractal) measure.

The measure most familiar to us is the probability measure. To understand scale-invariance in its terms, consider a discrete probability distribution given by

$$P(x) = \exp(-\mu)\, \mu^x \, / \, x!$$

This is a Poisson distribution. Here μ is the average value of the random variable x. The sum of k such variables will have a mean of $k\mu$. Hence, let

$$P_k(x) = \exp(-k\mu)\, (k\mu)^x / x!$$

and using Stirling approximation, according to which

$$x! \approx (2\pi)^{0.5} \, x^{x+0.5} \exp(-x), \qquad x > 0$$

we get

$$k^{-1} \ln [k \, P_k(k\alpha)] = -\mu + \alpha \ln (\mu e/\alpha) = \sigma_e(\alpha)$$

which is independent of k and dependent only on μ, apart from α, which is what scale-invariance is all about.

As another example, take a continuous random variable with a density function

$$p(x) = \exp(-x) \, x^{\beta-1} / \Gamma(\beta),$$

which is called a gamma density function. It has a property that the sum of two gamma-distributed random variables with parameters β_1 and β_2 is a gamma random variable with parameter $\beta = \beta_1 + \beta_2$. Hence, let

$$p_k(x) = \exp(-x) \, x^{k\beta-1}/\Gamma(k\beta)$$

Again, using Stirling approximation, we get

$$k^{-1} \ln [k \, p_k(k\alpha)] = \beta \ln (\alpha/\beta) - \alpha + \beta = \sigma_e(\alpha)$$

which is independent of k and dependent only on β, apart from α, as required by scale-invariance.

There are three more distributions, viz. binomial, normal and Cauchy, which have such a property. The function $\sigma_e(\alpha)$ is a limit probability distribution function plotted on doubly logarithmic co-ordinates. Another function

$$f_e = \sigma_e(\alpha) + 1$$

is also popular. The αs such that $f(\alpha) > 0$ are said to be manifest. Then the multifractal is said to be a pseudomultinomial. The other αs are said to be latent.

Examples

Omori's (1894) law in seismology can now be recognised as the first fractal law in seismology. According to it the rate of the aftershock events decays as the inverse power law in elapsed time after the main shock. It is worth rejoicing when we are in the fractal mood that this law is more than a century old.

The Gutenberg-Richter (1944) relation in seismology, which is more celebrated, states that

$$\log N_m = a - bm$$

where N_m is the number of earthquakes of magnitude exceeding m in unit time interval falling in a region under study. The range of b is 0.9 ± 0.1. This fractal relation is more than 50 years old. The *G-R* relation is supposed to be equivalent to (Aki, 1981)

$$N_m = k \, A^{-D/2}$$

where k is a constant and A is the rupture area on the fault that led to the earthquake. This is a fractal relation. It is based on the assumption that the fault has a fractal geometry.

The relation between D and b is

$$D = 3b/\delta$$

where δ is a constant depending on the relative durations of the seismic source and the time constant of the recording system. For crystalline rocks δ is taken to be 3. For most earthquake studies it is believed to be around 1.5. For subduction zones between 100-700 km depth the value suggested for it is 2.4. Correspondingly, the values of D would be b, $2b$ and $1.25b$ respectively.

Huxley (1932) dealt with the problem of different growths of parts of an organism. Let A and B be characteristic lengths of two parts. Then a simple model could be

$$A' = \alpha AG, \, B' = \beta BG,$$

where G is a time-dependent factor, characteristic of the whole organism, and prime denotes differentiation with respect to time. Then

$$d(\log A)/dt = k \, d(\log B)/dt$$

with $k = \alpha/\beta$. This leads to

$$A = bB^k$$

which is called Huxley's law of allometric growth (Lord and Wilson, 1986), and would now be recognised as a fractal law. Generation of biological form in response to physical forces has been fascinatingly dealt with by Thompson (1917).

GENERATION OF MULTIFRACTALS

Gips (1975) and Stiny (1975) used shape grammars to generate a variety of two-dimensional geometrical patterns. Let u, v, etc. denote shapes. Let Su, Sv, etc. be shapes similar to them respectively, wherein similarity transformation includes Euclidean transformation or dilatation. A typical shape rule (u, v) operating on shape w is (Lord and Wilson, 1986, pp. 136-143)

$$(u, v)w = w - Su + Sv$$

That is, if $Su \in w$, it is replaced by Sv in w. A simple example is shown in Fig. 1(A), wherein u is a straight-line segment and v is obtained by replacing the middle one-third segment in it by an angular indentation. Stiny (1975) and Gips (1975) use markers (here *s) which are not part of the shape and do not appear in the final shape, but indicate locations where some prescribed procedure is followed. In the above rule, the substitution also changes the number of markers. At the *, which is a marker, a straight line develops a symmetric angular indentation as shown. Starting with an equilateral triangle, this single-rule grammar will develop a Sierpinski triangle. Thus, if a rule replaces a shape by a shape with finer details, geometrical fractals can be generated by substitution shape grammars. The role of recursion can be clearly seen in this generation. The straight line has two sides. Therefore, markers can appear on either side or on both sides. Starting with a square, the fractal in Fig. 2 can be obtained by a shape grammar. Try to put markers (not shown in the figure) as that fractal recursively develops.

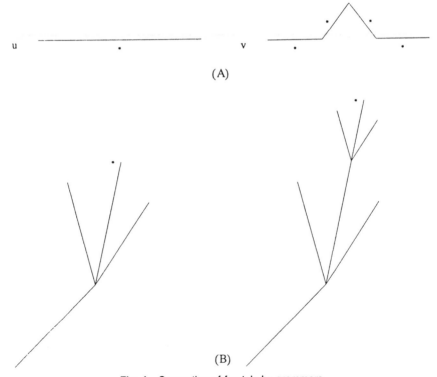

Fig. 1 Generation of fractals by grammars

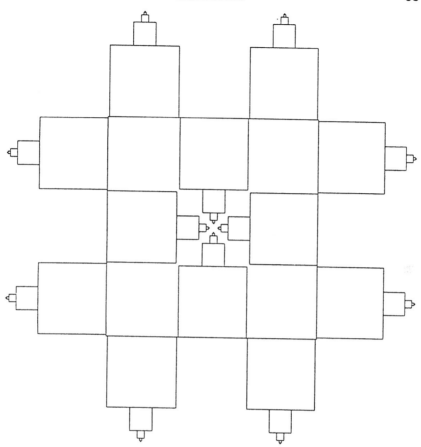

Fig. 2 A fractal generated by a shape grammar.

There are, of course, many other types of grammars that can generate fractals. For example, Stevens (1976) defines a parallel shape grammar, which can be exemplified in Fig. 1(B). Patterns in which branching takes place recursively can be generated by such a grammar. Rivers and tributaries, fractures and dendritic connections can so be simulated.

Another method for obtaining multifractals is described by Barnsley (1988) and Berger (1993). It uses an affine transformation

$$x_{n+1} = A\, x_n + b$$

in which **x** is a column vector, giving the geometrical co-ordinates of a point, n and $(n + 1)$ are consecutive points (as generated) on the fractal, **A** is a matrix and **b** is a column vector. The idea is to let **A** and **b** be randomly chosen. In actual implementation, they are chosen from a small number (even just two in many cases) of fixed possibilities, with prescribed probabilities. Six fractals thus obtained are shown in Figs. 3-8. Two types of self-similarity can be observed in them. A particular pattern repeats with close similarity at a particular scale at a number of places, and it similarly repeats at a number of different scales also. The theory of this method can be found in Berger (1993), but it is really amasing that simple probabilistically recursive affine transformations can lead to beautiful self-affine patterns. The generation of multifractals as above is in the nature of a forward problem. But an inverse or coding problem is that of finding the number of affine transformations that must be used, the probabilities with which they must be chosen and the estimation of **A** and **b** in those transformations, given the multifractal, which may have arisen in some physical study. If this problem is satisfactorily solved, these few parameters constitute an adequate description of the multifractals of interest. Then as a next step, the classification of multifractals can be performed in their terms. Barnsley (1988) has investigated these issues. The problem can be more difficult in some real applications. Frequently, we do not *see* multifractals in physical Euclidean spaces, but in metaphorical spaces. The empirical observations are in the form of a time-series or a point process. Then we visualise a space in which the phenomenon is a fractal. The dimension of that space is itself a mathematical construct. Then the coding problem is to be solved in that space.

Fig. 3   A multifractal obtained by random affine transformations

Fig. 4 A multifractal obtained by random affine transformations

Fig. 5 A multifractal obtained by random affine transformation

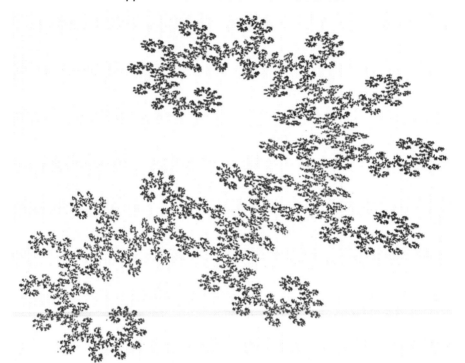

Fig.6 A multifractal obtained by random affine transformations

Fig.7 A multifractal obtained by random affine transformations

Fig.8 A multifractal obtained by random affine transformations

IMPORTANT SUGGESTION

Acceptance that the notions of fractals and multifractals are relevant in earth sciences should lead to revision of the methods of data collection (Turcotte, 1989).

ACKNOWLEDGEMENT

I am grateful to Shri V.M. Maru for intellectual interaction, to Dr. V.P. Dimri for inviting me to write for this volume and to Director, NGRI for material and motivational support.

REFERENCES

Aki K. 1981. A probabilistic synthesis of precursory phenomena. In: *Earthquake Prediction*, pp. 566-574. Simpson D.W. and Richards P.G. (eds.). American Geophysical Union, Wash., DC.

Barnsley MF. 1988. *Fractals Everywhere*. Acad. Press, NY.

Barton CC. 1995. Fractal analysis of scaling and spatial clustering of fractures. In: *Fractals in Earth Sciences*, pp. 141-178. Barton C.C. and La Pointe P.R. (eds.). Plenum Press, NY.

Bellman RE. 1961. *Adaptive Control Processes*. Princeton Univ. Press, Princeton.

Berger MA. 1993. *An Introduction to Probability and Stochastic Processes*. Springer-Verlag, NY, pp. 139-172.

Brown SR. 1995. Measuring the dimension of self-affine fractals: example of rough surfaces. In: *Fractals in Earth Sciences*, pp. 77-87. Barton C.C. and La Pointe P.R. (eds.). Plenum Press, NY.

Carrion P. 1987. *Inverse Problems and Tomography in Acoustics and Seismology.* Penn Publ. Co., Atlanta, pp. 176-181.

Erkman AM and Stephanou HE. 1990. Information fractals for evidential pattern classification, *IEEE Trans.*, SMC-20: 1103-1114.

Frisch U and Parisi G, 1985. Fully developed turbulence and intermittency. In: *Turbulence and Predictability in Geophysical Fluid Dynamics and Climate Dynamics*, p. 84. Ghil M. (ed.). North Holland, Amsterdam.

Gips J. 1975. *Shape Grammars and Their Uses.* Birkhäuser Verlag, Basel.

Gutenberg B and Richter CF. 1944. Frequency of earthquakes in California. *Bull. Seismol. Soc. America*, **34**: 185-188.

Hartmann WK. 1969. Terrestrial, lunar, and interplanetary rock fragmentation. *Icarus*, **10**: 201-213.

Huxley J. 1932. *Problems of Relative Growth*, Methuen, NY.

Klein F. 1903. Vergleichende Betrachtungen über neaere geometrische Forschungen. *Math. Annalen*, 43.

Lord EA and Wilson CB. 1986. *The Mathematical Description of Shape and Form.* Ellis Horwood Publ., Chichester, 260 pp.

Lorenz EN. 1993. *The Essence of Chaos.* Univ. of Washington Press, Seattle, 227 pp.

Mandelbrot BB. 1977. *The Fractal Geometry of Nature.* Freeman, San Francisco.

Mandelbrot BB. 1989. Multifractal measures, especially for geophysicists. In: *Fractals in Geophysics*, pp. 5-42. Scholz C.H. and Mandelbrot B.B. (eds.). Birkhäuser Verlag, Basel.

Moharir PS. 1992. *Pattern-Recognition Transforms.* Research Studies Press, Taunton, pp. 38-39.

Omori F. 1894. On aftershocks of earthquakes. *J. Coll. Sci. Imperial Univ. Tokyo*, **7**: 111-200.

Power WL and Tullis TE. 1995. Review of the fractal character of natural fault surfaces with implications for friction and the evolution of fault zones. In: *Fractals in Earth Sciences*, pp. 89-105, Barton C.C. and La Pointe PR. (eds.). Plenum Press, NY.

Pruess SA, 1995. Some remarks on the numerical estimation of fractal dimension. In: *Fractals in Earth Sciences*, pp. 65-75. Barton C.C. and La Pointe P.R. (eds.). Plenum Press, NY.

Ruellè D. 1990. Deterministic chaos: the science and the fiction. *Proc. Royal Soc. London*, **A247**: 241-248.

Sammis CG and Steacy SJ. 1995. Fractal fragmentation in crustal shear zones. In: *Fractals in Earth Sciences*, pp. 179-204. Barton C.C. and La Pointe P.R. (eds.). Plenum Press, NY.

Scholz CH and Aviles CA. 1986. In: *Earthquake Source Mechanisms.* Das S., Boatwright J. and Scholz C.H. (eds.). American Geophysical Union, Wash., DC.

Stevens PS. 1976. *Patterns in Nature.* Penguin Books, Tharmondsworth.

Stiny G. 1975. *Pictorial and Formal Aspects of Shape and Shape Grammars.* Birkhäuser Verlag, Basel.

Thompson DW. 1917. *On Growth and Form.* Cambridge Univ. Press.

Turcotte DL. 1989. Fractals in geology and geophysics. In: *Fractals in Geophysics.* pp. 171-196. Scholz C.H. and Mandelbrot B.B. (eds.). Birkhäuser Verlag, Basel.

Turcotte DL and Huang J. 1995. Fractal distributions in geology, scale-invariance, and deterministic chaos. In: *Fractals in Earth Sciences*, pp. 1-140. Barton C.C. and La Pointe P.R. (eds.) Plenum Press, NY.

PROCESSING (MULTI)FRACTAL
DATA STRINGS

Rita Singh*

INTRODUCTION

Every data string carries information. Both numerical and intelligent (symbolic) techniques of data analysis translate this information into different *forms* which may yield better insights into the process which the data string represents. So far, we have been translating this information into Euclidean or Riemannian forms—straight-lines and curves in Euclidean spaces and topologies in Riemannian spaces. Since the principles of geometry in these spaces are well established for a subset of possible forms, and since these forms are precisely measurable in terms of simple geometrical units, we have been able to quantify this information and draw meaningful inferences about the generating mechanism.

Recently, mathematicians have established the principles of a new geometry called *fractal geometry,* through which forms of intricate and infinite complexity in Riemannian spaces can be precisely described and measured. These complex geometrical structures are called *fractals.* In a measure-theoretic sense, every complex form of seemingly random topology may be constructed or represented through a set of superimposed or intertwined fractals. Such intertwined fractals are called *multifractals.* Fractal geometry is thus the final or complete geometry, in terms of which

* School of Computer Science, Carnegie - Mellon University 5000, Forbes Avenue, Pittsburgh, PA 15213, USA.

almost all real or hypothetical structures may be described. Since such structures are now mathematically tractable, it is possible to think of translating the information present in data strings into them in order to gain newer insights into the processes represented.

Translation of information present in data strings into fractal geometrical forms is called *fractal data analysis* and the process of drawing parametric inferences from these forms is called *fractal data interpretation*. A very important point to note here is that in order to interpret data through fractal forms, *it is not necessary* that the data itself be fractal, or the outcome of a fractal generator. All data strings, subject to certain universal constraints, may be convertible into fractal forms and interpretable by fractal techniques. Some data strings, on the other hand, are *themselves* fractals. They may be generated by fractal stochastic, deterministic-chaotic or hyperchaotic processes. Fractal data strings do not require auxiliary fractal analysis. They may be directly interpreted.

(Multi)fractals: lies that tell the truth

A fractal object is a geometrical structure which is everywhere continuous but nowhere differentiable. A fractal *set* is an infinite set with finite measure. Such a structure exists at theoretically infinite resolutions, and at any resolution its physical appearance is the same. Geometrically, the only qualitative difference between a fractal and a multifractal is that a multifractal appears to be equally complex, but *different* at different resolutions.

(Multi)fractals are thus *lies that tell the truth*, to borrow a phrase from Picasso. At a given resolution, a (multi)fractal is a structure which is *present* over some space. When one looks closely at a small region over which it is *present*, however, it may simply vanish and may give place to a structure as complex as the overall structure at the earlier (lower) resolution. In a fractal, a point that is, is not and a point that is not, may be. The cycle continues *ad infinitum* and yet, at every resolution, the individual fractal pattern and appearance are distinctive.

Fractality and information

Multifractal data analysis converts information into a fractal geometrical structure in a hypothetical space. A fractal structure is a completely ordered structure. Does this mean that we can relate entropy—which is the conventional measure of information, or loss of it, or disorder—to a fractal's representation of information in terms of its precise and intrinsic order? By fractal data analysis, do we represent entropy as an order-parameter? Fractal data processing is indeed the reflection of entropy into order. The following sections discuss order-parameters or *scaling exponents*, information and entropy leading to the order-parameter called *information dimension,* and attempt to bridge and clarify some basic issues in fractal geometry and fractal data processing.

The significance of a scaling exponent

In Fig. 1, the approximate length $L(r)$ of the upper curve is given by the product $N \times r$, where N is the number of straight-line segments of length r needed to step along the curve from one end to the other and r is the length of each segment or the *unit*. From the figure, it is clear that as r becomes smaller and smaller, $L(r)$ approaches a limit which is the actual length L of the curve. In measuring an area too, as r becomes smaller and smaller, the area measured tends to a constant A. This is not true for a fractal. If the curve shown in the upper part of the figure were a fractal, $L(r)$ would diverge to infinity in the limit $r \to 0$.

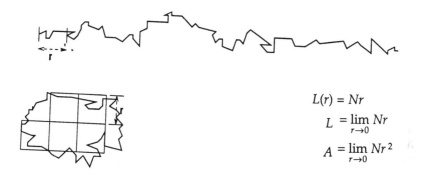

$$L(r) = Nr$$
$$L = \lim_{r \to 0} Nr$$
$$A = \lim_{r \to 0} Nr^2$$

Fig.1 Dependence of measured quantities on units of measurement

This can be clearly seen through Fig. 2, which shows the construction of a Koch curve and its use in generating the Koch fractal. At each stage, every straight-line segment in the previous stage is segmented into three equal parts and the sides of an equilateral triangle are traced on the middle part with the base removed.

At each stage or resolution n, then, the length of a straight-line segment becomes $\frac{1}{3^n}$ times the length of the straight-line segment r_0, while the length of the curve measured in terms of the length of the unit at that stage works out to be 4^n. Thus it would seem that as the unit of measurement becomes smaller and smaller, the length measured becomes larger and larger and diverges to infinity in the limit

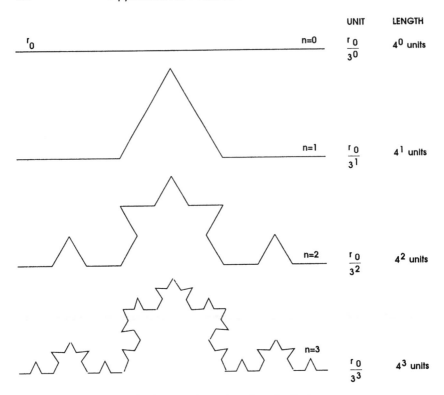

	UNIT	LENGTH
r_0 n=0	$\dfrac{r_0}{3^0}$	4^0 units
n=1	$\dfrac{r_0}{3^1}$	4^1 units
n=2	$\dfrac{r_0}{3^2}$	4^2 units
n=3	$\dfrac{r_0}{3^3}$	4^3 units

Fig. 2 Construction and scaling of a Koch curve

$$L(r) = \lim_{r \to 0} N \times r = \infty \tag{1}$$

This divergence to infinity can be controlled by raising the unit of measurement r to a power D, so that the length measured is constant in the limit $r \to 0$, i.e.

$$L(r) = \lim_{r \to 0} N \times r^D = \kappa \tag{2}$$

where κ is a finite constant. Assuming that $\kappa = 1$ in the above equation, the exponent D for the Koch curve can then be calculated as

$$D = \frac{\log N}{\log\left(\dfrac{1}{r}\right)} = \frac{\log 4^n}{\log\left(\dfrac{3^n}{r_0}\right)} \tag{3}$$

Assuming r_0 to be a unit length, this works out to be

$$D = \frac{\log 4}{\log 3} = 1.26... \tag{4}$$

for the Koch curve. D is thus a **scaling exponent**, since it controls and describes the behaviour of the measurement functions as the curve evolves at all scales or resolutions. For the Koch curve, it is clear from eqn. (2) that if $D > 1.26..$, the product $N \times r^D$ tends to infinity in the limit $r \to 0$, while if $D < 1.26..$, the product tends to 0. D is thus a critical value, and is therefore called the **critical exponent**.

Raising the unit r to a power D is not the only way to limit the behaviour of different measures on a fractal to a finite value. There may be other ways of capturing the scaling properties of a fractal. In general, all such numbers which are critical in the description of the geometry of a fractal at all resolutions are called **dimensions**. A fractal may therefore have an infinite set of dimensions. Figure 3 shows the construction of a Cantor set, which is a fractal *set* and not a fractal *curve* like the Koch curve. A fractal curve is a continuous but non-differentiable curve. A fractal set, like the Cantor set, is a collection of infinite points of zero total length. A fractal set is also called a *point set*. The scaling exponent for the Cantor set may be calculated by taking into account the fact that at any given resolution n, the unit of measurement scales as $r_0/3^n$, while the length measured in terms of this unit scales as 2^n. The critical exponent—also called the *Hausdorff dimension* if calculated in this manner—works out to be

$$D = \frac{\log 2^n}{\log 3^n} = \frac{\log 2}{\log 3} = 0.63... \tag{5}$$

Generally, $D > 1$ for fractal curves and $D < 1$ for fractal sets. When $D = 1$, however, the curve is not a fractal, since the scaling exponent is an integer and not *fractional*.

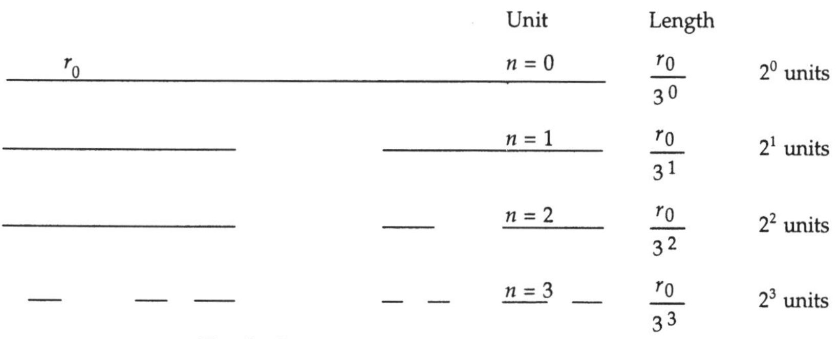

	Unit	Length	
r_0	$n = 0$	$\dfrac{r_0}{3^0}$	2^0 units
	$n = 1$	$\dfrac{r_0}{3^1}$	2^1 units
	$n = 2$	$\dfrac{r_0}{3^2}$	2^2 units
	$n = 3$	$\dfrac{r_0}{3^3}$	2^3 units

Fig. 3 Construction and scaling of a Cantor set

CREATING AND UNDERSTANDING FRACTALS WITH
VARIABLE SCALING EXPONENTS

Fractals can be easily created. For example, Fig. 4 shows the outline of a sample lecture which could be given by a speaker following the general trend in this paper. This is itself a fractal. Let us call this the *talk-fractal*.

At the centre of this Figure is the main topic (resolution=0), which at the next resolution (resolution=1) splits up into two topics, namely, **what we need to know about a fractal** and **what we need to know about data strings**. At the next resolution (resolution=2), these subtopics again split into two subtopics each and the process continues. If actual words were not taken into account and if the sample lecture continued indefinitely, magnification of any portion of the structure shown would result in a structure similar to the overall structure in Fig. 4. This illustrates the property of **scale-invariance** of a fractal. The fractal has a similar (self-similar) structure at all scales. In other words, it is similar to itself at all resolutions.

EXAMPLES
ONE FRACTAL = INFINEITE COMPLEXITY: HOW?
GENERATING INFINITE COMPLEXITY ON THE COMPUTER

WHAT WE NEED TO KNOW ABOUT A FRACTAL

EXAMPLES : FRACTAL CURVE

ONE FRACTAL = ONE NUMBER: HOW?

FRACTAL DATA SETS
CAN WE BUILD FRACTALS : CANTOR SET AND THIS FRACTAL
MULTIFRACTALS

PROCESSING (MULTI)FRACTAL DATA STRINGS

SHANNON'S MEASURE OF INFORMATION

ENTROPY, ORDER AND INFORMATION

COMPLEX SIGNALS (IN GEOPHYSICS)
COMPLEXITY
SOME USEFUL HINTS ON COMPLEXITY

WHAT WE NEED TO KNOW ABOUT DATA STRINGS

OPTICAL NOISE
NOISE: THE UNWANTED SIGNAL
OTHER TYPES OF NOISE

Fig. 4   The talk fractal

Figure 5 shows the overall schematic structure of the talk-fractal shown in Fig. 4, where each topic is represented by a line. Figure 6 illustrates how the talk-fractal can be collapsed into a point set and also mapped into a bush. The upper part of this figure is a Cantor-like point set, in which each straight-line segment represents a topic. Beginning with a single topic at $n = 0$, there is a recursive splitting of each topic into two subtopics (along with the geometrical scaling of the segments). A little thought would show that at each resolution n of this talk-fractal, the unit of measurement is $1/5^n$, while the length measured in terms of this unit is $3(2^{n-1}) + \sum_{k=0}^{n-2} 2^k 5^{n-1-k}$, where negative indices are inapplicable.

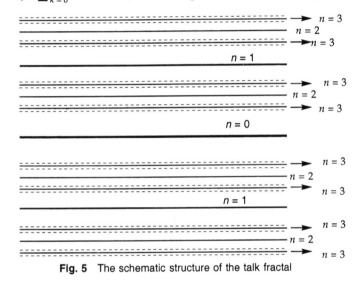

Fig. 5 The schematic structure of the talk fractal

Calculation of the scaling exponent for this fractal leads to the expression

$$D = \frac{\log\left[3(2^{n-1}) + \sum_{k=0}^{n-2} 2^k 5^{n-1-k}\right]}{\log(5^n)} \qquad (6)$$

Clearly, the scaling exponent D is dependent on the resolution n and changes with it. When a fractal has resolution-dependent scaling exponents, it is called a multifractal. Figure 7 shows a plot of D versus n. Since D is clearly a *measure* plotted against the *supporting set n*, the plot of D versus n is a *spectrum*. It is thus one of the many possible multifractal spectra.

A fractal can actually be collapsed into other fractals using simple rules formulated out of analogy. As an example, the lower part of Fig. 6 shows how the talk-fractal may alternatively be mapped into a fractal structure which has the appearance of a bush. Every topic is represented

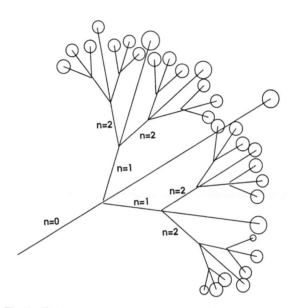

r_0 $n = 0$

$n = 1$

$n = 2$

$n = 3$

Fig. 6 The talk fractal collapsed in to a point set and a bush

by a branch in this bush. The circles at the terminal positions of this bush have simply been drawn to provide a better visual effect.

GENERATING FRACTALS

Generating a fractal with a given scaling exponent

Figure 8 outlines the process of generation of a fractal curve with $D = 1.18...$ The number $1.18...$ is expressible as $\log (10)/\log (7)$. Thus one needs to generate a fractal whose unit scales as $1/7^n$ at the nth resolution and whose measure length at each resolution scales as 10^n. This can be done by generating a Koch-like curve, which divides into seven segments instead of three (unlike the Koch fractal), and raising the sides of an equilateral triangle over three segments as shown in the Figure. The process is continued recursively, generating a fractal of dimension $1.18...$

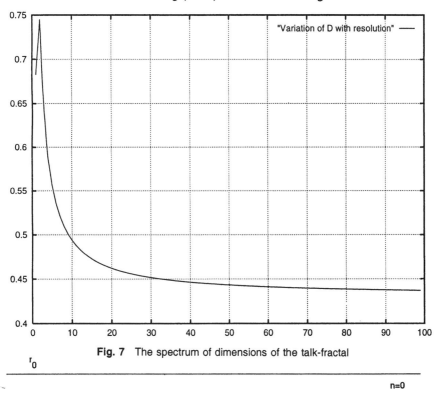

Fig. 7 The spectrum of dimensions of the talk-fractal

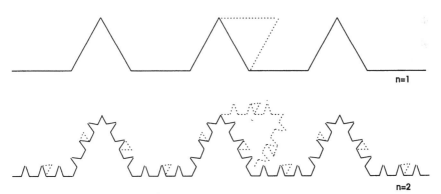

Fig. 8 Generating a fractal of dimension 1.18...

Similarly, a fractal of dimension log (11)/log (7)=1.23... may be generated by distorting any one of the sides of the equilateral triangles into an equilateral triangle as shown by the dotted lines in the Figure. One thus generates two equilateral triangles and one quadrilateral (with the base removed).

Generating fractals through parallel rewriting systems : turtle graphics

Fractals are of two kinds:
1) those generated by growth processes,
2) those generated by solutions of non-linear equations.

Fractals generated by growth processes are called L-system fractals after Aristid Lindenmeyer, a biologist who invented the formalism for their computergraphic generation in the context of natural growth systems such as plants. L-systems are alternatively called *'parallel rewriting systems'*. An L-system is a logical framework for the generation of fractals and can be implemented as turtle-graphic codes. A turtle-graphic code begins with an axiom, which is the first or the basic structure along which an imaginary turtle may be imagined to move, as shown in Fig. 9.

heh...heh...

Axiom ———— **Production rules** ———→ Fractal

Fig. 9 The simple essentials of turtle graphics

Figure 10 shows the formal framework for writing an L-system using the Koch curve as an example. In Fig. 10, the turtle begins with an axiom F and follows a production-rule-defined path F+F– –F+F on cue to each appearance of the axiom in the previous stage. In other words, the axiom is iterated through or replaced by one or more production rules during the construction stages. The signs + and – denote left and right turns respectively while the parameter δ is the angle through which these turns are taken. On successive iterations of such a production rule, the Koch curve is generated.

Figures 11-13 show the L-systems for the Cantor set, the talk-fractal collapsed into a point set and into a bush (the brackets in these figures denote branching) and some examples of L-systems generated by the production rules shown in Fig. 13. L-systems which have more than one production rule are called Iterated Function Systems (IFS). These are shown in Figs. 12-13. Figure 12 is a good example of what are generally called *iterated two-function deterministic multifractals*. If an IFS iterates through its production rules in a predetermined or fixed manner, it is called a deterministic IFS. If there are two or more choices to be made at each implementation of the production rules, i.e., if the production rules are selected as per a probability distribution, the L-system is called a stochastic IFS. The geometrical structure of many natural systems can be

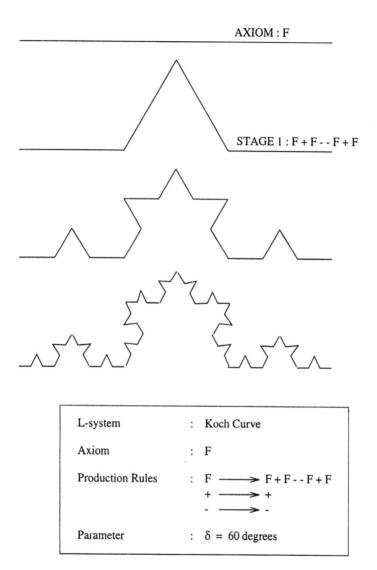

Fig. 10 Construction of the Koch fractal using turtle graphics

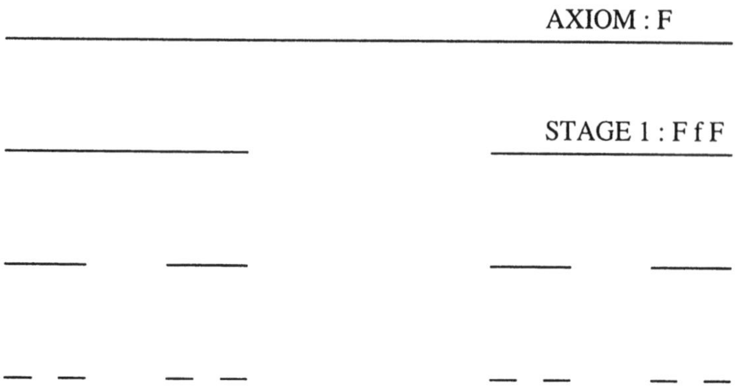

AXIOM : F

STAGE 1 : F f F

L-system	:	Cantor set
Axiom	:	F
Production Rules	:	F ⟶ F f F
		f ⟶ f f f

Axiom : F
Stage 1 : F f F
Stage 2 : F f F fff F f F
Stage 3 : F f F fff F f F fff fff fff F f F fff F f F

Fig. 11 Cantor set as an L-system

simulated by L-systems, which constitute a powerful formalism in fractal computergraphics. IFS-generated fractals are also called networked fractals.

DIGITAL TO FRACTAL (D/F) CONVERSION: PRELIMINARIES

This can be done in two ways:
1. From the given data string, find the fractal dimension and map a fractal of the same dimension onto a data string. *Note that the data string does* not *have the same dimension, but is a* mapping *of a fractal.* The new fractal data string can then be used for further studies.
2. Find the dimensionality of the embedding space by estimating the correlation function and reconstruct the strange attractor (or trajectory) within the embedding space.

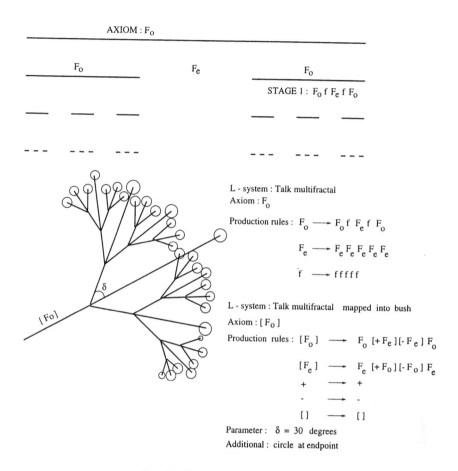

AXIOM : F_o

F_o F_e F_o

STAGE 1 : F_o f F_e f F_o

L - system : Talk multifractal
Axiom : F_o

Production rules : $F_o \longrightarrow F_o$ f F_e f F_o

$F_e \longrightarrow F_e F_e F_e F_e F_e$

f $\longrightarrow$ f f f f f

L - system : Talk multifractal   mapped into bush
Axiom : [F_o]

Production rules : $[F_o] \longrightarrow F_o$ [+ F_e][- F_e] F_o

$[F_e] \longrightarrow F_e$ [+ F_o][- F_o] F_e

$+ \longrightarrow +$

$- \longrightarrow -$

$[] \longrightarrow []$

Parameter : $\delta = 30$ degrees
Additional : circle at endpoint

Fig. 12   Cantor set as an L-system

Constructing (multi)fractal data strings from point sets and L-systems

When the dimension of a data string is found to be less than, 1, point sets can be used to construct a fractal of the same dimension and these new point sets can then be mapped onto a data string. The new data string is a fractal. Similarly, when the given data string has a dimension which lies between 1 and 2, fractal curves can be used to construct the new fractal data strings. Figure 14 shows an example. In this Figure, a data string is constructed out of a Cantor set by positioning an impluse of magnitude equal to the length of a segment at a position corresponding to the midpoint of the segment. The process is terminated with some

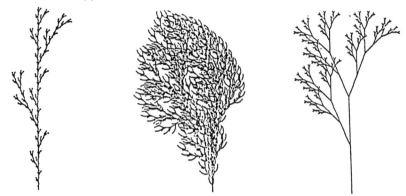

L-system: Weedlike Plant I   Production rules: $F \rightarrow FF$
Axiom: F $B \rightarrow F[+B]F[-B] + B$
Production rules: $F \rightarrow F[+F]F[-F]F$ Parameter: $\delta = 20$ degrees
Parameter: $\delta = 25.7$ degrees

axiom F, production $F \rightarrow FF + [+F - F - F] - [-F + F + F]$
and angle $\delta = 25$ degrees.

L-system: Stochastic Weedlike Plant III
Axiom: F
Production rules: $F \rightarrow F[+F]F[-F]F$ (probability 1/3)
                        $F \rightarrow F[+F]F$ (probability 1/3)
                        $F \rightarrow F[-F]F$ (probability 1/3)
Parameter: $\delta = 25.7$ degrees

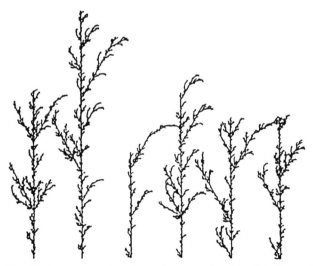

Fig. 13 Some examples of growth system fractals. This figure shows both deterministic and stochastic iterated function systems along with their L-system formalism.

resolution of the Cantor set. The data string thus generated is equisampled by choosing a sampling interval Δ equal to the minimum spacing between two samples on the data string. The samples are then projected on a new string using a closest-neighbour-roundup criterion (which is flexible) giving the required data string. Where the closest neighbours are absent, zeros are substituted in the final fractal data string. In Fig. 14, the *natural sampling* rate of the data string is $\dfrac{1}{\Delta}$ = 27 Hz.

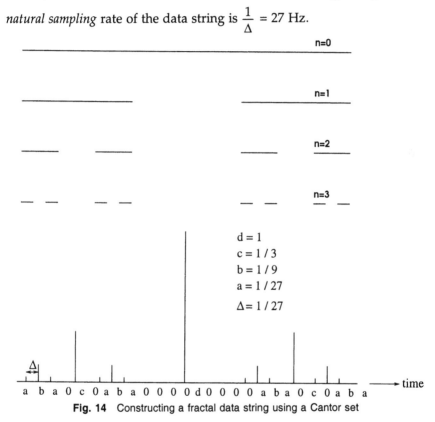

Fig. 14   Constructing a fractal data string using a Cantor set

A fractal of any dimension $D = \log (m)/\log (n) < 1$ can be obtained by constructing a Cantor-like set which divides into n equal parts and has m countable ones, both scaling uniformly with resolution. On the other hand, a fractal of any dimension $D = \log (m)/\log (n) > 1$ and be obtained by constructing a Koch-like curve as in Fig. 10. At each resolution, the unit divides into n parts and the trace of the curve distorts to give m countable ones. The L-string for such a construction at any stage (see, for example, stage 3 of Fig. 10) can be translated into numbers in a suitable way, say by replacing every symbol in the L-string by a specific number or an operation on a specific number or sets of numbers. So to project a fractal of $D = \log(m)/\log(n) > 1$, one *first* generates an L-system starting from an axiom F and proceeding up to a certain resolution. The turtle-string of the L-system is then *translated* into numbers, and operations are prescibed over entire substrings. The turtle-string thus gets converted into a digital data string, which is a fractal. The string is then equisampled and the required fractal data string is obtained. The choices can be infinitely many—and by the same token such a data string would be an extremely non-standard string. Methods which may give more well-defined and specific data strings are yet to be devised. It must be noted here that the processing of L-system strings as symbolic data sets directly (using the rules of symbolic computation) would ultimately be a more desirable recourse.

Thus D/F conversions of these kinds can be done by the computer using simple L-system rules and a set of operations. The L-systems used may be deterministic or stochastic, singluar or iterated functions systems, depending on the specific requirements.

Constructing (multi)fractal objects using attractor reconstruction techniques (ART)

ARTs are borrowed from the chaos theory. Given a time-series, these techniques reconstruct the trajectory of the underlying strange-attractor (or phase orbit) within an estimated embedding space. A strange attractor is a fractal geometrical object, having multifractal measures, within which the trajectory described by the spatiotemporal evolution of the underlying (generating) physical system is chaotic.

The dimensionality of the embedding space is first estimated. Let this be N. As an example, let this be $N = 3$. Figure 15 pictures the basic procedure involved in the reconstruction of the trajectory from the given data string. The procedure can broadly be divided into two sets of operations:

1. Estimating the linearized flow map from the data string.
2. Reconstruction of local trajectory segments from the local state—transition matrix estimated.

The linearised flow map around a vector is called the flow map of the tangent space and is described by a state transition matrix which applies to the local evolution of the trajectory to be reconstructed. It is periodically renormalised.

Figure 15 actually shows the first set of operations. Since $N = 3$, three-component vectors are first reconstructed from given data string using some specific time lag τ (t in the Figure). The maximum possible lag values which could be used are

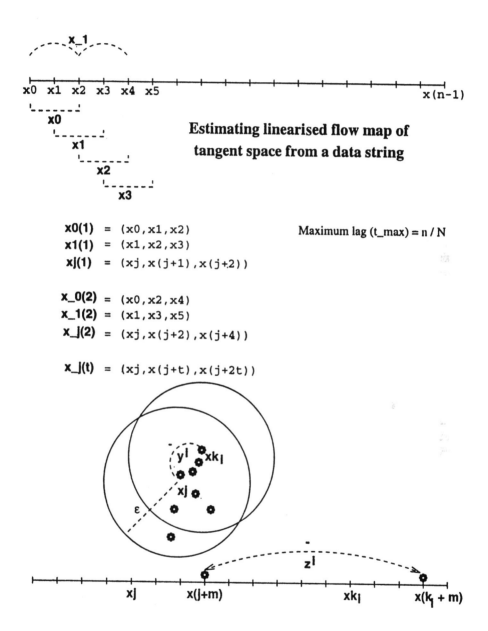

Estimating linearised flow map of tangent space from a data string

x0(1) = (x0,x1,x2)
x1(1) = (x1,x2,x3)
xj(1) = (xj,x(j+1),x(j+2))

Maximum lag (t_max) = n / N

x_0(2) = (x0,x2,x4)
x_1(2) = (x1,x3,x5)
x_J(2) = (xj,x(j+2),x(j+4))

x_J(t) = (xj,x(j+t),x(j+2t))

Fig. 15 Estimating the linearised flow map of tangent space from a data string: the first steps

$$\tau = \frac{n}{N} \tag{7}$$

Let the string

$$\{x\} = \{x_0, x_1, x_2, ...\} \tag{8}$$

represent the vector string thus computed. The components of the vector x_j are

$$x_j = x(t_0 + (j + 1)\, \Delta t) \tag{9}$$

where t denotes time. A set of points

$$\{x_i\}\ i = 0, 1, 2...., (N - 1) \tag{10}$$

are found within an ε neighbourhood of the vector x_j and the set of difference vectors within this neighbourhood is denoted by the set $\{y_i\}$, i.e.,

$$\{y_i\} = \{x_{ki} - x_j \mid \| x_{ki} - x_j \| \leq \varepsilon\} \tag{11}$$

The norm is either a Euclidean norm or an L_1-norm (for computationally faster implementations). The Euclidean norm is given by

$$\|w\| = \left(w_1^2 + w_2^2 + w_3^2 + ... + w_N^2\right)^{1/2} \tag{12}$$

where the ws are the components of the vectors $\mathbf{w}$. After an elapse of time $\tau = m\Delta t$, the vectors x_j and $x_{ki.}$ evolve as

$$x_j \rightarrow x_{j+m} \tag{13}$$

$$x_{ki} \rightarrow x_{ki+m} \tag{14}$$

A set $\mathbf{z}^i$ of new difference vectors is then computed as

$$\{\mathbf{z}^i\} = \{x_{ki+m} - x_{j+m} \mid \|x_{ki} - x_j\| \leq \varepsilon\} \tag{15}$$

From this follows the matrix equation

$$\mathbf{z}^i = \mathbf{A}_j\, \mathbf{y}_i \tag{16}$$

The matrix $\mathbf{A}_j$ is an approximation of the linearised flow map at x_j. The matrix $\mathbf{A}_j$ can be estimated from the set of vectors $\mathbf{y}^i$ and $\mathbf{z}^i$ by a mean square error minimising procedure:

$$\min S_{\mathbf{A}_j} = \min {}_{\mathbf{A}_j} \frac{1}{N} \sum_{i=1}^{N} \|z^i - \mathbf{A}_j\, \mathbf{y}_i\| \tag{17}$$

This minimises the average of the squared error norm between the vectors $\mathbf{z}^i$ and $\mathbf{A}_j\mathbf{y}^i$. Denoting the $(k, l)^{\text{th}}$ component of the matrix $\mathbf{A}_j$ by $a_{k,l}(j)$, we obtain $N \times N$ equations

$$\frac{\partial S}{\partial a_{kl}(j)} = 0 \tag{18}$$

which give us an expression for A_j, namely

$$A_j V = C \tag{19}$$

where

$$(V)_{kl} = \frac{1}{N} \sum_{i=1}^{N-1} y^{ik} y^{il} \tag{20}$$

$$(C)_{kl} = \frac{1}{N} \sum_{i=1}^{N-1} z^{ik} y^{il} \tag{21}$$

V and C are $N \times N$ covariance matrices, and y^{ik} and z^{ik} are the kth components of the vectors y^i and z^i respectively. A_j is then used as a local state-transition matrix to generate some p vectors starting from the vector x_j. The whole procedure is recursively followed for every vector x_j initially computed. The p vectors generated in this manner from all points in the data series are plotted to give the strange attractor—the required fractal object in phases space.

INFORMATION, ENTROPY AND INFORMATION DIMENSION

The given data string can be called a *source* **S**;

$$S = \{x_0, x_1,...., x_{n-1}\} \tag{22}$$

The data points then constitute the *source alphabet*. Let $P(x_i)_{i=0,1,2,...,(n-1)}$ be a probability distribution on the source **S**. $P(x_i)$ is the probability that on random sampling of the source, the element x_i would be chosen. Before the source is sampled, there is a certain amount of uncertainty associated with the outcome. After sampling one gains some information about the source. The concepts of uncertainty and information are thus related. For example, in a sourse if $P(x_0) = 1$ and $P(x_i) = 0$ for all $i > 0$, then the element x_0 is always chosen—and random sampling gives absolutely no new information about the source. On the other hand, the information gained is maximum when all outcomes are equally likely, i.e.,

$$P(x_i) = \frac{1}{n} \qquad i = 0, 1,..., (n-1)$$

Let us define a function $H(P(x_0), P(x_1), ..., P(x_{n-1}))$ or $H(P_0, P_1, ..., P_{n-1})$ which measures this uncertainty or information. One can logically form three important relations for this function:

1. If all probabilities are equal, then the greater the number of samples, the greater the information, i.e.

$$H\left(\frac{1}{n},\frac{1}{n},\ldots,\frac{1}{n}\right) < H\left(\frac{1}{n+1},\frac{1}{n+1},\ldots,\frac{1}{n+1}\right) \tag{23}$$

2. $H(P_0, P_1, \ldots, P_{n-1})$ is defined and continuous for all $P_0, P_1, \ldots, P_{n-1}$, $0 \le P_i$, $\Sigma P_i = 1$

3. Partitioning the elements of the source into non-empty disjoint blocks $\mathbf{B}_0, \mathbf{B}_1, \ldots, \mathbf{B}_{k-1}$, where $|\mathbf{B}_i| = b_i$ is the number of elements in each block and $\Sigma b_i = n$, we compute the conditional probility $P(x_j|\mathbf{B}_i)$. It is obvious that

$$P(\mathbf{B}_i) = \frac{b_i}{n} \tag{24}$$

The conditional probability that one would choose an element x_i which would additionally be in the block $\mathbf{B}_u$ is given by

$$P(x_j | \mathbf{B}_i) = 0 \qquad \text{if} \quad i \ne u \tag{25}$$

$$= \frac{1}{b_u} \qquad \text{if} \quad i = u$$

The probability of choosing x_j in $\mathbf{B}_u$ is thus

$$P(x_j) = \sum_{i=0}^{n-1} P(x_j|\mathbf{B}_i)P(B_i) = \frac{1}{b_u} \times \frac{b_u}{n} = \frac{1}{n} \tag{26}$$

so the probability of choosing x_j is the same under these (blocking) conditions—as if we had chosen directly from $\mathbf{S}$ with equal probabilities. The uncertainty in choosing one of the blocks is

$$H\left(\frac{b_1}{n},\frac{b_2}{n},\ldots,\frac{b_k}{n}\right) \tag{27}$$

and the uncertainty in choosing an element in the block is

$$\sum_{i=0}^{k-1} P(\mathbf{B}_i) \; (\text{uncertainty in choosing from } \mathbf{B}_i) = \sum_{i=0}^{k-1} \frac{b_i}{n} H\left(\frac{1}{b_i},\frac{1}{b_i},\ldots\frac{1}{b_i}\right) \tag{28}$$

Thus one arrives at the relation

$$H\left(\frac{1}{n},\frac{1}{n},\ldots\frac{1}{n}\right) = H\left(\frac{b_1}{n},\frac{b_2}{n},\ldots\frac{b_k}{n}\right) + \sum_{i=0}^{k-1} \frac{b_i}{n} H\left(\frac{1}{b_i},\frac{1}{b_i},\ldots,\frac{1}{b_i}\right) \tag{29}$$

One of the functions that can satisfy conditions 1, 2 and 3 above is the function

$$\mathbf{H}_b\,(P_0,\,P_1,\,...,\,P_{n-1}) = -\sum_{i=0}^{n-1} P_i \log_b P_i \tag{30}$$

This function is called the entropy function. Its units (units of information) depend on the base b of the logarithm. For a $\log_2$ expression, information is measured in units of *bits*. When the base is e (natural logarithm), information is measured in units of *nats*. $\mathbf{H}_b$ is called the b-ary entropy of the probability distribution.

To compute the entropy of a given data string, the samples are first grouped into histograms of prescribed bin-widths. The area under the histogram is then normalised to give the probability distribution. As an example, Fig. 16 shows the probability distribution obtained from a segment of digital data corresponding to one of the channels of a stereo recording (audio) of the recent Jabalpur earthquake (shown in Fig. 17). The entropy calculated form this distribution is approximately 3.9 nats.

The scaling exponent which describes the scaling of this entropy with bin-width is called an *information dimension*. If h_i denotes the bin-width of the histogram, then the Shannon information dimension is:

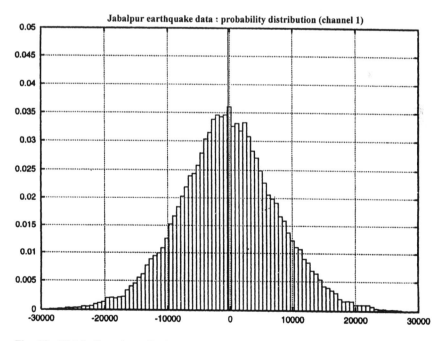

Fig. 16 Distribution of amplitudes on a section of one channel of the sound data corresponding to the Jabalpur earthquake (22 May 1997)

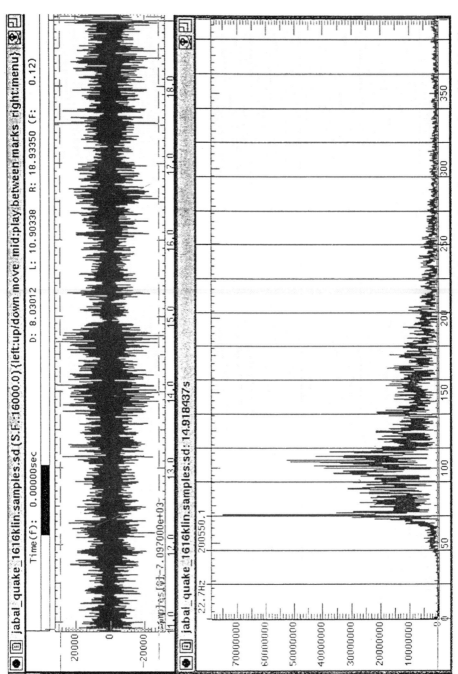

Fig. 17 One channel of a stereo recording (audio) of the 1997 Jabalpur earthquake. The lower part shows the power spectrum of the data stting

$$D_{Shannon} = \frac{\sum_{i=1}^{N} P_i \, \log P_i}{\sum_{i=1}^{N} P_i \, \log h_i} \tag{31}$$

where N is the number of bins used

REFERENCES

Farmer JD 1982. Dimension, fractal measure and chaotic dynamics. *In: Evolution of Order and Chaos.* H. Haken (ed.). Springer-Verlag, Berlin/New York.

Meakin P, Coniglio A, Stanley HE. and Witten TA. 1986. Scaling properties for the surfaces of fractal and nonfractal objects: an infinite hierarchy of critical exponents. *Phys. Rev.,* **A34:** 3325-3340.

Peitgen HO and Saupe D. 1988. *The Science of Fractal Images.* Springer-Verlag, New York.

Schroeder MR. 1990. *Number Theory in Science and Communication, with Applications in Cryptography, Physics, Digital Information, Computing and Self-Similarity* (2nd enlarged ed.). Springer-Verlag, Berlin/New York.

Schroeder M. 1991. *Fractal, Chaos, Power Laws: Minutes from an Infinite Paradise.* W.H. Freeman and Company, San Francisco.

Stanley HE and Meakin P. 1988. Multifractal phenomena in physics and chemistry. *Nature,* **335:** 405-409.

Tang C and Bak P. 1988. Critical exponents and scaling relations for self-organized critical phenomena. *Phys. Rev: Lett.,* **60:** 2347-2350.

Weyl H. 1981. *Symmetric.* Brikhäuser Verlag, Basel.

FRACTALS AND GEOLOGY

R.K. Sukhtankar*

INTRODUCTION

Mandelbrot coined the term 'fractal', mainly to describe the similarity of spatial patterns of some phenomena on many different scales. Generally, the approach is to describe any natural form by simple regular figures of finite length, up to a certain approximation. Geology is no exception; let it be any natural feature in the field of sedimentology, geomorphology, structural geology and the like. However, there is a limit to the length of a natural feature, as the yardstick for measurement is shortened to zero.

FRACTAL DIMENSION

According to Mandelbrot, any feature dimensionally lies between the familiar topological dimensions. Thus, for a straight line, it is $L1$; whereas a plane surface has dimension $L2$, while our familiar world has dimension $L3$. All these powers of dimension of length, i.e., 1, 2, 3..., are integers.

Any self-similar line, e.g. a river or a stream, flows on the plane, such as progressively to fill up more and more of it. As a straight line has a length dimension of 1, a river that flows so as to completely cover the plane, must have the length dimension, 2. From this observation, it

* Department of Geology, Shivaji University Centre for P.G. Studies, Solapur 413 003, India.

follows that a fractal line will have a dimension lying between 1 and 2, that would be a non-integral dimension. This has led Mandelbrot to formally define fractal as follows:

'A fractal.......... defined as a set for which the Hausdorff-Besicovitch dimension strictly exceeds the topological dimension.'

To understand the concept of fractal and fractal dimension, select any suitable stream course of reasonable length from the topographic map, of length say L. This length is measured by selecting any yardstick, say l_1 ; and with the reduction in yardstick, say as $l_2, l_3,....$, total length of a stream is measured. It is observed that there is variation in L, as the distance in yardstick is reduced. If the graph is plotted by using linear or arithmetic scale, with estimates of L on the ordinate, and the yardstick l on the abscissa, it may result as shown in Fig. 1(a). The relationship, however is not exactly linear. Instead, if the graph is plotted on logarithmic scale, it appears as a straight line and a linear relationship is revealed; this is termed the Richardson plot, as shown in Fig. 1(b).

The inferences from such a linear relationship can be drawn as follows:

(i) The slope of the straight line is somehow related to the sinuosity of the feature measured, being steeper for a more sinuous feature.

(ii) It is evident from both the plots that the more one reduces the yardstick of measurement and thus increases the accuracy of the estimate of the total length, the longer the line one seems to get.

(iii) The degree of sinuosity involved at each step is much the same, both the large- and small-scale features are similar geometrically except for their scale.

This is the property of self-similarity.

Thus, from the logarithmic plot in Fig. 1(b), the Hausdorff-Basicovitch dimension or fractal dimension can be estimated from the slope of the line of the feature.

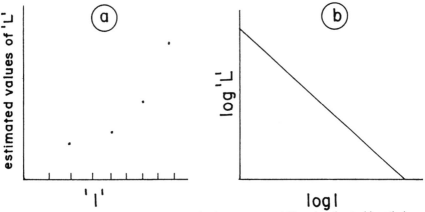

Fig. 1   A fractal relation between unit of measurement (*l*) and estimated length *L*

The logarithmic plot shown in Fig. 1(b) is known as the Richardson plot; as stated earlier, the same is used to determine the fractal value or the fractual dimension.

APPLICATION IN COASTLINE PROBLEMS

Practically, the idea of a fractal can best be understood from the coastline problem, as suggested by Richardson. He mainly used this diagram to compare the coastlines of the world (Fig. 2). As stated earlier, the gentler the slope of the best-fit regression line, the simpler the configuration of the coastline. Also, if these lines of different coasts are parallel or nearly parallel to each other, it can be inferred that such coastlines have identical configuration. Mandelbrot has further extended the idea of such plots to investigate the amount of convolutions and also to specify the actual coastline length.

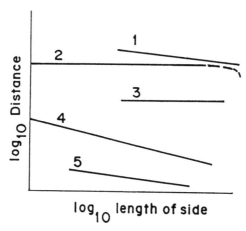

Fig. 2 Estimated distance versus length of measuring side for (1) Australian coast, (2) Circle, (3) South African coast, (4) west coast of Britain, (5) Portugese land frontier (after Richardson, 1961).

The Richardson plot, i.e., log P (perimeter of the estimated length of an object) versus log s (step length) usually results in an inverse linear function, from which the Hausdorff-Besicovitch dimension or fractal dimension, D, is given as, $D = b + 1$, where b is the slope of best-fit regression line.

The value of D can fall anywhere between 1 and 2. If $D = 1$, then it is a Euclidean Figure or at least the morphology is essentially simple and can be functionally characterised by axial ratio. As the value of D approaches to two, the line becomes increasingly plane-filling.

APPLICATION IN SEDIMENTOLOGY

In the field of sedimentology, quantification of the morphology of a sedimentary particle has been attempted by various methods and interpretations and inferences there from are derived. For the purpose, three axes mutually at a right angle to each other of a sedimentary particle are considered and axial ratios are worked out. However, consideration of three axes is possible only in the case of very coarse particles, such as a boulder, cobble or pebble and for finer sizes, e.g. gravel or sand, the particle's projected outline in two dimensions on a plane is considered and the ratio of longer and shorter axes of the same are taken to describe the particle's morphology. In other words, the shape, or sphericity of the particle's morphological envelope is considered in terms of similarity to a standard reference solid, e.g. sphere or ellipsoid. However, it has been noted that in most cases, particles defy morphological descriptions in terms of such axial ratios and form indices. This is evident from Fig. 3, in which both the particles, though having the same axial ratio, differ in their edge outline texture. It is therefore clear that the problem of reducing two-dimensional variations of the particle outline onto a linear scale has not been satisfactorily and fully solved.

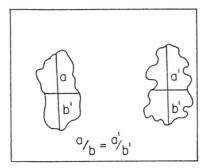

Fig. 3 · Fractality of the outlines of sedimentary particles

As shown in Fig. 3, the presence of jagged edge-morphology and/or crenellate edge morphology on the particle outline cannot be described or found a problem in description. Such a problem can be solved, however, with the help of fractal value and fractal dimension.

As stated earlier, the fractal value or fractal dimension relates to the best-fit regression line in the Richardson plot, thus identifying a single fractal, with an overall single dimension, which has been termed D_1 by Orford and Whalley (1983).

However, Kaye (1978) indicated that the behaviour of certain closed loops was not as clear cut as that of the Richardson plot for coastlines. He observed that the linear form of the Richardson plot was not typical for fine particles. He suggested that instead of fitting a single line, two lines were more appropriate, which he termed the structural (large scale and shape) and textural (edge-shaped) fractal components (Fig. 4).

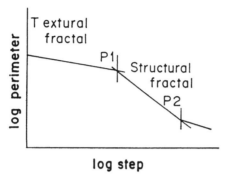

log step

Fig. 4 Bifractality of sedimentary particles; microscale edge and shape of particles reflect as textural and structural fractal respectively (after Orford and Whalley, 1983)

These constitute a departure from the concept of 'self-similarity' and therefore, are appropriately referred to as pseudofractals for outlines, as described in the Richardson plot. Each component in the plot, therefore, has its own fractal dimension.

Orford and Whalley (1983) have extended the idea of Kaye and have used three-line fittings for some plots and have classified and described highly indented particle outlines, by just textural and structural components and also by a single (total) fractal line.

Sukhtankar and Pandit (1995) carried out fractal dimension studies of the coastal Quaternary sediments from Maharashtra, which had been characterised by both structural and textural fractal elements. Selection of such sediment particles has an advantage, aṣ the original morphological characteristics of these sedimentary particles have not been obliterated or otherwise modified by post-depositional processes such as compaction, lithification and diagenesis.

On the basis of fractal values, Orford and Whalley (1983) classified sediment particles into three different groups, having morphological characteristics of their own. Considering the relative abundance of groups in the sediments analysed, textural maturity of sediments can be discussed.

CONCLUSION

The concept of fractal and fractal dimension is rather new and is gradually finding place for application in Earth Sciences. Moreover, it is an additional tool and can be considered complementary to the methods used earlier for analysis.

REFERENCES

Kaye BH. 1978. Specification of the ruggedness and/or structure of a fine particle profile by its fractal dimension. *Powder Technol*. **21**: 1-16.

Orford JD and Whalley WB. 1983. The use of fractal dimension to quantify the morphology of irregular-shaped particles. *Sedimentology*, **30**: 655-668.

Richardson LF. 1961. The problem of cantiguity. An appendix to the statistics of Deadly Quarrels. *Gen. Sys.*, **6**: 139-187.

Sukhtankar RK and Pandit SJ. 1995. Fraction dimension studies of sediment particles. *Bull. ONGC Limited*, **32**: 71-77.

CRUSTAL FRACTAL MAGNETISATION

V.P. Dimri*

INTRODUCTION

The spectral analysis method (Spector and Grant, 1970) has been widely applied to gravity and magnetic data, mainly in order to estimate the thickness of a sedimentary basin. The popular method makes an assumption that the source distribution is random. The assumption of random distribution makes the estimation of source parameters simpler. However, many borehole data around the globe have shown that density or susceptibility distribution is not random. Rather it is fractal. Hence there is a need to introduce fractal distribution of source instead of random distribution of source in the formulation of gravity and magnetic problems.

FRACTAL DISTRIBUTION

Density, susceptibility distribution and reflectivity sequence of many boreholes around the globe have been investigated. The power spectral density of density, susceptibility distributions and reflectivity sequences of some of the boreholes are shown in Fig. 1. The power spectrum of susceptibility distribution from 4 km depth of the German Continental

* National Geophysical Research Institute, Hyderabad 500 007, India.

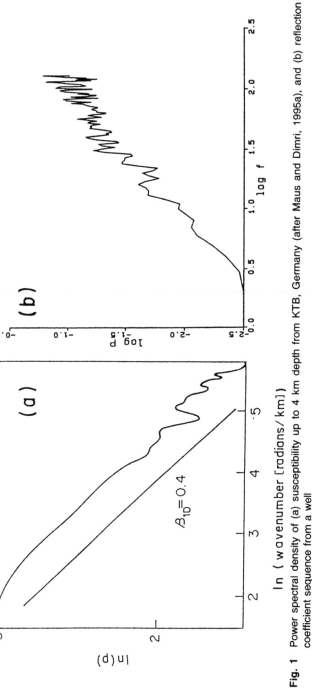

Fig. 1 Power spectral density of (a) susceptibility up to 4 km depth from KTB, Germany (after Maus and Dimri, 1995a), and (b) reflection coefficient sequence from a well

Deep Drilling Project (KTB) was computed by Maus and Dimri (1995a). From these Figures it is clearly seen that the power spectrum is not flat, rather it follows some scaling laws.

APPLICATIONS TO POTENTIAL FIELDS

Interpretation of potential field data can either be done in space (time) or in the frequency domain. The results from the two approaches are the same as can be seen below:

For linear gravity or magnetic system, the gravity or magnetic field is expressed as

$$H(x) = G(x) * M(x) \tag{1}$$

where $H(x)$ is the observed gravity or magnetic field; $G(x)$ the Green function for a source geometry; $M(x)$ is the density or the susceptibility distribution and * indicates convolution.

Transforming eqn (1) in the frequency domain

$$H(f) = G(f) M(f) \tag{2}$$

where $H(f)$, $G(f)$ and $M(f)$ are the Fourier transform of $H(x)$, $G(x)$ and $M(x)$ respectively.

The amount of information in both equations or domains is the same, however, the mathematical operation in eqn (2) is multiplication instead of convolution as in eqn (1). The choice of domain of operation depends on the problem on hand such as in the following cases:

Case: 1

Consider that the observed field is free from a corrupt noise. So, the density or susceptibility function can be found from eqn (2) as

$$M(f) = \frac{H(f) G(f)^*}{|G(f)|^2} \tag{3}$$

where $G(f)^*$ is complex conjugate of $G(f)$.

Case: 2

Let density or susceptibility distribution be random; hence its power spectral density is white. So, it becomes simpler to find a relation between the power spectrum of the observed fields and the source geometry. This situation has been utilised to estimate source geometry from the power spectrum of the observed fields. The first paper in this direction appeared from Spector and Grant (1970), closely followed by Naidu (1970). Thereafter many variants of the Spector and Grant technique for interpretation of gravity and magnetic data in the frequency domain appeared for different one- and two-dimensional geological structures. The method became very popular (see Hildenbrand *et al.*, 1993 and references therein). The method has been widely used for estimation of sedimentary thickness, thickness of volcanic suits etc. from gravity and aeromagnetic data as given below:

The power spectrum of an observed aeromagnetic field at height, h, $P_h(r)$ is related to the power spectrum of the top of magnetic surfaces $P_0(r)$ as

$$P_h(r) = e^{-2hr} P_0(r) \tag{4}$$

where r is wave number.

It is a well-known upward and downward continuation relationship of potential fields in a medium free of sources.

Let us assume the power spectrum due to magnetic sources, $P_0(r)$ is constant. Taking the logarithm of eqn (4) yields

$$\ln P_h(r) = -2hr + c \tag{5}$$

where c is constant.

Equation (5) is a straight-line equation whose slope is $-2h$, if a graph between logarithm of power spectrum of observed field and wave number (r) is plotted. The subsequent straight-line slopes give different magnetic interfaces (layers) as shown in Fig. 2.

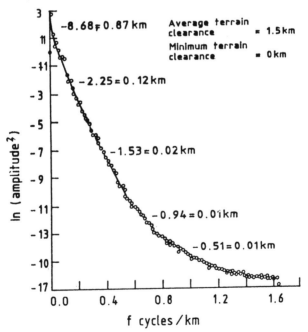

Fig. 2 The different slopes of the power spectrum provide thickness of the magnetic layers (after Connard *et al.*, 1983)

As mentioned earlier, the assumption of power spectrum due to sources as a constant is not true in nature. Let us see the power spectrum of aeromagnetic data from an area in Hawaii in a single logarithm scale in Fig. 3, which is not flat but decays exponentially. Converting Fig. 3 into a double logarithm scale as shown in Fig. 4 provides a straight-line. Mathematically, Fig. 4 means:

$$\ln P_0(r) = -\gamma \ln r + k$$

$$P_0(r) = k \, r^{-\gamma} \tag{6}$$

where k and γ are constants.

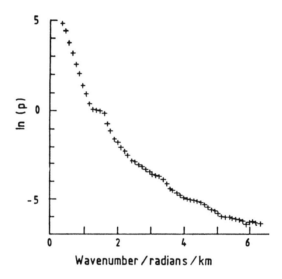

Fig. 3 The power spectral density of aeromagnetic data of Hawaii plotted in semilog scales (after Hildenbrand *et al.*, 1993)

The power spectrum defined by eqn (6) is the power spectrum of a scaling noise (Mandelbrot, 1983). The constant γ is called the scaling exponent of the scaling noise. Pilkington and Todoeschuck (1993) and Maus and Dimri (1994; 1995a, b; 1996) have obtained the value of scaling exponent for different regions of the world from gravity and aeromagnetic data.

Combining eqns (4) and (6) yields

$$P_h(r) = k \, e^{-2hr} \, r^{-\gamma} \tag{7}$$

Equation (7) can be used to estimate the thickness of a sedimentary basin and the thickness of volcanic basalts from gravity and aeromagnetic data provided the constants k and γ are predetermined. Maus and

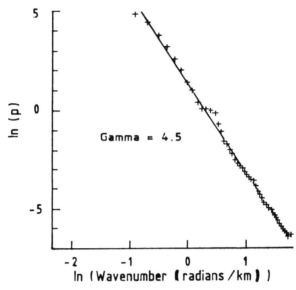

Fig. 4 The power spectral density of Fig. 3 in double logarithmic scale
(after Maus and Dimri, 1995b)

Dimri (1996) have applied the new method to several field studies and results are very encouraging. Currently, we are applying this method to gravity and magnetic data to Nagaur-Jhalawar and Jaipur-Raipur transects, India, in order to model the substructure of the area. The work of Maus and Dimri has also been cited by Pilkington and Todoeschuck, 1995; Fedi *et al.*, 1997; Zhou and Thybo, 1998 and several others.

The new technique needs the value of γ, as seen from eqn (7). In general, the value of the scaling exponent due to sources is not known. Maus and Dimri (1995a) derived a relation between the scaling exponent between gravity and magnetic fields and their sources which is given as:

$$\gamma_g = \beta_{density} + 1 \qquad (8)$$

for gravity, and

$$\gamma_m = \beta_{susceptibility} - 1 \qquad (9)$$

for magnetic fields.

Here γ_g, γ_m, $\beta_{density}$ and $\beta_{susceptibility}$ are scaling exponents of gravity, magnetic fields, density and susceptibility distribution respectively. The advantage of eqns (8) and (9) is that, if one scaling exponent is known the other can be estimated.

CONCLUSIONS

The concept of fractals has been introduced in the interpretation of gravity and magnetic data. A new method has been developed to estimate the thickness of a sedimentary basin or thickness of volcanic traps whose density or susceptibility distribution is not random but follows natural scaling laws. The technique has been applied to several field studies and now is being applied to gravity and magnetic data from Nagaur-Jhalawar and Jaipur-Raipur transacts, India, in order to model the subsurface for a realistic geology of the area.

Such observations of scaling distributions have been verified from susceptibility data obtained up to 4 km depth from the German Continental Deep Drilling Project (KTB) and other regions of the world. The finding that the crustal magnetisation is indeed fractal may be able to redefine Curie depth (Maus *et al.*, 1997) and detectability limits of geophysical surveys (Dimri, 1998).

REFERENCES

Connard G, Couch R and Gemperle M. 1983. Analysis of aeromagnetic measurements from Cascade range in Central Oregon. *Geophysics*, **48**: 376-390.

Dimri VP. 1998. Fractal behaviour and detectability limits of geophysical surveys. *Geophysics*, **63**: 1943-1946.

Fedi M, Quarta T and Santis A.D. 1997. Inherent power-law behaviour of magnetic field power spectra from a Spector and Grant ensemble. *Geophysics*, **62**: 1143-1150.

Hildenbrand TG, Rosenbaum JG and Kauahikaua JP. 1993. Aeromagnetic of the Island of Hawaii. *J. Geophy. Res.*, **98**: 4099-4119.

Mandelbrot BB. 1983. *The Fractal Geometry of Nature*, WH Freeman, NY.

Maus S and Dimri VP. 1994. Scaling properties of potential fields due to scaling sources. *Geophy. Res. Lett.*, **21**: 891-894.

Maus S and Dimri VP. 1995a. Potential field power spectrum inversion for scaling geology. *J. Geophys. Res.*, **100**: 12605-12616.

Maus S and Dimri VP. 1995b. Basin depth estimation using scaling properties of potential fields. *J AEG*, **16**: 131-139.

Maus S and Dimri VP. 1996. Depth estimation from the scaling power spectrum of potential fields ? *Geophys. J. Int.*, **124**: 113-120.

Maus S, Gordon D and Fairhead D. 1997. Curie-temperature depth estimation using self-similar magnetisation model. *Geophys. J. Int.*, **129**: 163-168.

Naidu PS. 1970. Statistical structure of aeromagnetic field. *Geophysics*, **35**: 279-292.

Pilkington M and Todoeschuck JP. 1993. Fractal magnetization of continental crust. *Geophys. Res. Lett.*, **20**: 627-630.

Pilkington M and Todoeschuck JP. 1995. Scaling nature of crustal susceptibility. *GRL*, **22**: 779-782.

Spector A and Grant FS. 1970. Statistical models for interpreting aeromagnetic data. *Geophysics*, **35**: 293-302.

Zhou S and Thybo H. 1998. Power spectral analysis of geomagnetic data and KTB susceptibility logs, and their implication for fractal behaviour of crustal magnetisation. *PAGEOPH*, **151**: 147-159.

FRACTALITY OF SEISMIC WAVE SIGNATURE—A MANDELBROT APPROACH

N.L. Mohan* and L. Anand Babu**

INTRODUCTION

Fractal studies pertaining to seismic wave signatures are, to the best of the authors' knowledge, at a primitive stage (Davis *et al.*, 1994; Scholz and Mandelbrot, 1989; Mohan, 1995; Mohan and Anand Babu, 1997(a) and (b)), and need serious attention, because the very nature of the earth is heterogeneous and hence falls under a non-linear regime. The wave shape and magnitude depend on several parameters, such as source energy, velocity, density, depth, topography, shape of the target, incident/reflected/diffracted/refracted angles.

We consider the seismic source pulse as $f(t)$, and the reflected/refracted wave form as $f(a(t - t_0))$, where the scale factor, $a = x' + i z'$ (Fig. 1) (Appendix A) and the corresponding Mellin-Fourier transforms (Appendix B) are given by (Mohan *et al.*, 1991; Mohan and Babu, 1992)

$$M_1(s) = \int_0^\infty F_1(w) w^{-is-1} dw; \quad \text{and} \quad M_2(s) = \int_0^\infty F_2(w/a) w^{-is-1} dw \tag{1}$$

* Centre of Exploration Geophysics, Osmania University, Hyderabad-500 007, India.
** Department of Mathematics, Osmania University, Hyderabad-500 007, India.

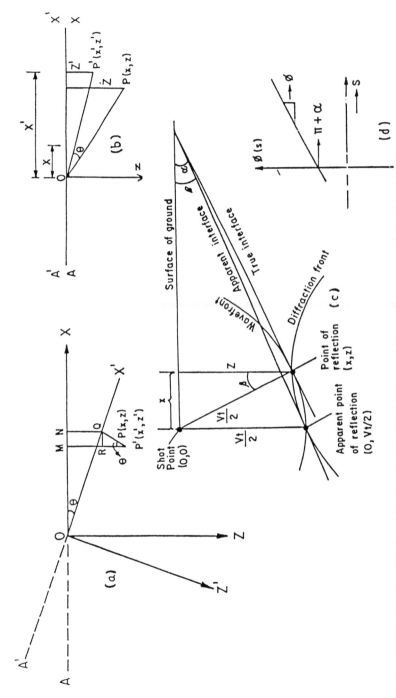

Fig. 1 (a) Keeping the origin 0 and depth point P(x, z) fixed, the co-ordinate axes are rotated clockwise by θ degrees;

(b) Alternarive to the system (a), keeping the axis AOX (coincides with axis A'OX') fixed, point P(x,z) is rotated to P'(x', z') in the anticlockwise direction by the angle θ degrees;

(c) Phase component of the Mellin-Fourier transform of the scaled analytic wavelet;

(d) Schematic diagram for migration.

$$M_2 (s) = a^{-is} M_2 (s) = \exp(-i \lambda s) M_1 (s)$$

$$(a = e^\lambda \text{ or } \lambda = \ln a) \tag{2}$$

where $F_1 (w)$ and $F_2 (w/a)$ are Fourier transformation of respective time functions $f(t)$ and $f(a(t - t_0))$ and accordingly the magnitude of the reflected waveform is written as $| M_1(s) | = | M_2(s) |$ (Fig. 2), that is by eliminating the scale factor, the reflected (or diffracted) wavelet virtually reduces to the magnitude of source pulse. It may be mentioned that the stretching operation is non-linear and so band limited signals will not have the same shape in the stretched domain. The band widths of the stretched wavelets will depend on wavelet position (Bickel, 1993). The shift property of the Fourier and the scale property of the Mellin transforms are given in Appendix C. Also, the Mellin power spectrum and the Mellin-Fourier transform of the scaled analytic Rayleigh wavelet are discussed in Appendixes D and E. Further, the phase component of the MFT of $f(a(t - t_0))$ is written as

$$P(s) = \arctan [-\text{Im}\{M(s)\}/\text{Re}\{M(s)\}] = -s(p+iq) \tag{3a}$$

where

$$p = \ln (\text{sqrt} (x'^2 + z'^2)) \text{ and } q = \arctan (-z'/x') \tag{3b}$$

TEST OF ANALYTIC

Here, $P(s)$ may be defined as the logarithmic function of complex analysis by the equation

$$\log Z = \log r + i \, \theta \tag{4}$$

where $r = | z |$ and $\theta = \arg z$. It is a multiple-value function, which is defined for all non-zero complex numbers and by generalisation, we write as

$$\log Z = \log r + i \, (\phi + 2\pi l), \qquad \text{where } l = \pm 1, \pm 2, \pm 3,... \tag{5}$$

It is evident that the real part remains the same for any value of l, and the imaginary part differs by multiples of integral 2π, and for $l = 0$

$$\log Z = \log r + i \, \phi \qquad r > 0; \; -\pi < \phi < \pi$$

We define the component functions as

$$u(r, \phi) = \log r \, ;$$

and
$$v(r, \phi) = \phi \tag{6}$$

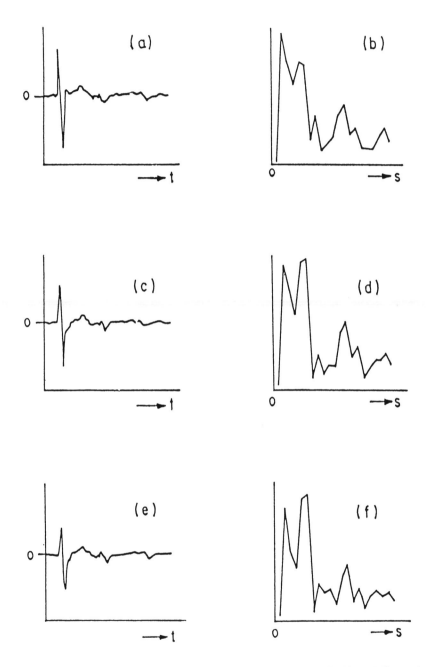

Fig. 2   (a), (c), and (e) are 10°, 20°, and 30° angular wavelets. (b), (d), and (f) are the normalised Mellin-Fourier spectra of the wavelets (a), (c), and (e) respectively

Each of the component functions, and hence log z, is continuous in the domain R and satisfies the Cauchy-Reimann equations (Appendix F) everywhere in the domain. If we restrict the value θ in eqn (4) so that $\alpha < \theta < \alpha + 2\pi$ for a fixed but arbitrary α, the function

$$\log Z = \log r + i\theta \qquad (r > 0; \alpha < \theta < \alpha + 2\pi)$$

is single valued and continuous throughout the domain. Further, if $w = \log z$, the range is horizontal strip $\alpha < \text{Im } \alpha < \alpha + 2\pi$ (Fig. 3). Hence log Z is analytic and also satisfies the polar form of the Cauchy-Reimann equations. Further, a branch of a multiple-value function f is any single-valued function F, which is analytic in some domain at each point Z, of which the value F(Z) is one of the values of f(Z). Accordingly, log Z defined at points in the domain $r > 0$, $-\pi < \phi < \pi$ constitutes a branch of the logarithmic function (eqn. (4)) and the branch is called the principal branch. Each point of the negative real axis $\phi = \pi$, as well as the origin, is a singular point of the principal branch log Z. The ray $\phi = \pi$ is called the branch cut for the principal branch and is either a line or a curve of singular points. The ray $\phi = \pi$ is a branch cut for the branch (eqn (4)) of the logarithmic function (Fig. 3). The singular point Z = 0, common to all branch cuts for that multiple-valued function, is called a branch point.

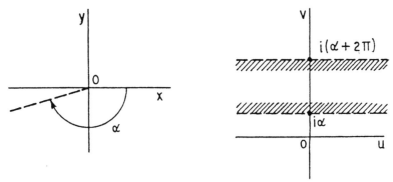

Fig. 3 The characteristic feature of the function log Z (= log r + i θ) is analytic in the domain r > 0; α < θ < α + 2π.

ANALYSIS

(a)Test of validity of λ

Differentiating eqn (2) with respect to the scale constant λ, ($a = e^\lambda$ or $\lambda = \ln a = \ln(x' + iz')$) and equating to zero we obtain

$$dM_2(s)/d\lambda = -is \cdot \exp(-is\lambda) \cdot M_1(s) = 0 \qquad (7)$$

or $\exp(is\lambda) = 0$ (since $M_1(s) \neq 0$)

i.e., $\exp(sq) \cdot [\cos(s\,p) - i\sin(s\,p)] = 0 + i0$ (8)

where p and q are defined in eqn (3b). Consequently, equating the real and imaginary components to zero, the respective resultants are:

$$x'^2 + z'^2 = e^{(2l+1)\pi/S} ; \quad \text{and} \quad x'^2 + z'^2 = e^{2l\pi/S} \qquad (9)$$

From the above equation it is evident that the expansion or contraction of circular wavefronts takes place for a fixed value of +1 and –1 respectively. Further, eqn (9) is recast as

$$l = (s\log(vt)/\pi) - 0.5; \quad \text{and} \quad l = s\log(vt)/\pi ; \qquad (10)$$

$$(x'^2 + z'^2 = (vt)^2)$$

respectively. Thus, it is clearly established that the expansion or compression factor of the seismic signature is quantifiable, is related to the velocity × time (= depth) and varies with respect to the value of s; and hence can be determined (Fig. 1(a))

(b) Determinability of λ

Considering the phase component of the Mellin-Fourier transform

$$P(s) = -s\lambda = -s\ln(x' + iz')$$

The scale factor of the seismic wave signature $|\lambda|$ can be determined as

$$|\lambda| = |P(s)/s| \qquad (11a)$$

That is $|P(s)/s|$ versus s is a straight line parallel to the abscissa. The width along the ordinate gives the scale factor or 'time stretch' of the seismic wave signature. Also, by differentiating $P(s)$ with respect to s the scale factor or time stretch factor (Fig. 1) can be determined as

$$|\lambda| = |dP(s)/ds| \qquad (11b)$$

Interestingly, from eqn (2) it is known that the magnitude of the M-F transform of the reflected/diffracted wave signature is equal to the source wave signature. Accordingly, the scale factor or time stretch factor can also be determined as

$$|\lambda| = |(1/s)\ln[M_1(s)/M_2(s)]| \qquad (11c)$$

Evidently, $|(1/s)\ln[M_1(s)/M_2(s)]|$ versus s yields the straight line parallel to the abscissa and the width along the ordinate is the scale or time stretch factor.

FORMULATION OF 'TIME STRETCH' FACTOR USING MANDELBROT APPROACH

We define the equation for the Mandelbrot set as (Stevens, 1993; Schuster, 1988),

$$P_n = P_{n-1}^2 + s \tag{12}$$

and the potential of the Mandelbrot set is given by

$$G(s) = \log |P_n(s)|/2^2 \tag{13}$$

Accordingly, at any point s the 'time stretch' factor, T, to the Mandelbrot set can be defined from the phase component of the MFT and its first derivative with respect to s (eqns (3) and (3a)),

$$|T| = G(s)/2 |G'(s)|$$

or

$$|T| = |P(s)| \log |P(s)|/2 |P'(s)|$$

$$= |(s/2) [\log s + \log(\mathrm{sqrt}\,(p^2 + q^2)]| \tag{14}$$

where

$$|P(s)| = |\lambda s|;$$

$$\log |P(s)| = \log |\lambda s|; \text{ and}$$

$$|P'(s)| = |\lambda|$$

The 'time stretch' estimate (Fig. 1(c) and (d)), similar to the distance estimate (Stevens, 1993), has two characteristics: (a) there is no possibility of reaching with an estimate that is inside the Mandelbrot set as it is always an underestimate, and (b) it refines as the time stretch from the point to the Mandelbrot set decreases (Stevens, 1993; Scholz and Mandelbrot, 1989).

LYAPUNOV EXPONENT—A CHARACTERISTIC FEATURE OF CHAOTIC MOTION

The basic ingredients of the Bernoulli shift property of the map are the stretching and the backfolding, a mechanism of generating deterministic chaos. In this context, from the phase component of the Mellin-Fourier transform of the reflected/diffracted wave signature (scaled waveform, eqn 11(b)), the Lyapunov exponent can be defined as:

$$|\lambda(s_0)| = \lim_{N\to\infty} (1/N) (1/s) \log |P'(s)| \tag{15}$$

Generalising for the seismic trace, the Lyapunov exponent is

$$|\lambda(s_0)| = \lim_{N\to\infty} (1/N) \sum_{i=0}^{N-1} (1/s_i) \log |P'(s_i)| \tag{16}$$

Further, the mean loss of information $(\overline{\Delta I})$ (associates with iterations with a linear map) can be defined as

$$\overline{\Delta I} = -\lim_{N\to\infty} (1/N) \sum_{i=0}^{N-1} (1/s_i) \log_2 |P'(s_i)| \tag{17}$$

Now the mean loss of information $(\overline{\Delta I})$ can be related to the Lyapunov exponent as

$$|\lambda(s_0)| = (\log 2) \cdot |\overline{\Delta I}| \tag{18}$$

The relation between the Lyapunov exponent and the loss of information is a first step towards characterisation of chaos in a co-ordinate invariant system.

EXAMPLES

(a) *Simulated example:* Here scaled wavelets are generated (Fig. 4) and the corresponding M-F transformation of the wavelets computed. From the phase component, the scale factor (or time stretch factor) of the wavelets are determined (Fig. 4) and it is evident that the scale factors (or time stretches) are constant with respect to the Mellin variable s.

(b) *Practical example:* The phase component of the M-F transformation of a recorded seismic trace evidently shows that the trace has three scale factors (time stretches, Fig. 5) and it is obvious that the recorded seismic signature contains three reflected events from different depth horizons. Here it may be noted that since the entire seismic trace is subjected to the M-F transformation, the scale (or time stretch) segment B (from A to B and B to C) is gently dropping from A to C. However, this needs further detailed study.

CONCLUSIONS

We have attempted to highlight certain important ramifications from a mathematical analysis of the seismic wave signature: (1) the scale factor can be defined as complex, which associates with offset distance, depth, velocity, angles of incident/reflected rays, and dip of geological target;

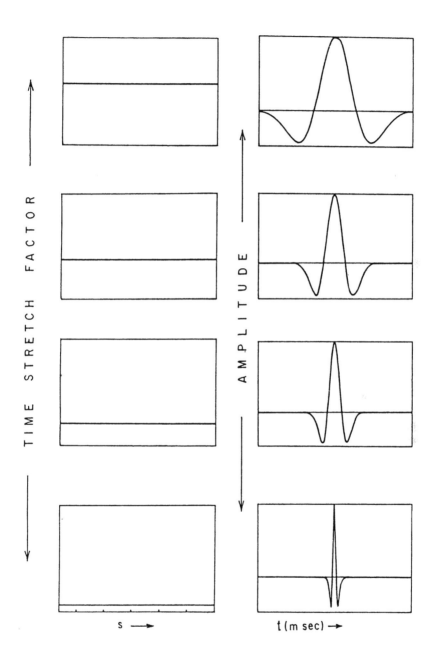

Fig. 4 Wavelets and corresponding time stretch factors (obtained from the phase component of the Mellin-Fourier transform)

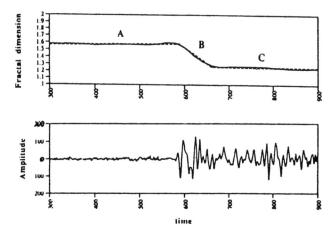

Fig. 5 A recorded seismic trace and its fractal (or time stretch) dimension. A, B, and C represent the time stretch factors of three consecutive seismic events embedded in the recorded trace

(2) by removing the scale factor, using the M-F transformation, the reflected waveform can be reduced to the original source pulse; (3) the phase component of M-F transformation is complex and satisfies the test of analyticity; (4) from $dM_2(s)/d\lambda = 0$, expansion and contraction of the wave signature can be explained; (5) it is shown that the scale factor is constant for any particular value of s; (6) using the Mandelbrot equation 'time stretch' can be defined and quantified; (7) using the time stretch factor, the Lyapunov exponent can be defined and (8) the relation between the Lyapunov exponent and mean loss of information is formulated and is a first step towards characterisation of chaos in a co-ordinate invariant system. We hope this study will provide deeper insights into the chaos ànd fractals of the elastic media of the earth.

ACKNOWLEDGEMENTS

We profusely thank the Department of Science and Technology, Government of India, New Delhi, for having sanctioned the financial grant under a research project.

REFERENCES

Bickel SH. 1993. Similarity and the inverse Q filter: the Pareto-Levy stretch. *Geophysics*, **58**(11): 1629-1633.

Churchill RV, Brown JW and Verhey RF. 1974. *Complex Variables and Applications*, McGraw Hill International Book Co., NY

Davis A, Marshak A and Wiscombe W. 1994. Wavelet-based multifractal analysis of non-stationary and/or intermittent geophysical signals. In: *Wavelets in Geophysics*. pp. 249-298. Efi Foufoula-Georgiou and Praveen Kumar (eds). Academic Press, Inc., New York.

Hubral P and Tygel M. 1989. Analysis of the Rayleigh pulse. *Geophysics* **54**(5): 654-658.

Mohan NL. 1995. Discussion on: "Pulse distortion in depth migration". *Geophysics*, **60**(5): 1948-1949.

Mohan NL and Anand Babu L. 1992. A new mathematical tool: Mellin-Fourier transform—An application in seismics, C-MMACS Intl. Symp. Mathematical Modelling and Scientific Computing, pp. 1-9.

Mohan NL and Anand Babu L. 1997(a), Fractality of seismic wave signature—A Mandelbrot approach. Paper presented at DST Workshop on "Applications of Fractals in Geophysics". NGRI, Hyderabad, India.

Mohan NL and Anand Babu L. 1997(b). Can travel time curve tunnel through chaotic regime. Paper presented at the DST Workshop on 'Application of Fractals in Geophysics', NGRI, Hyderabad, India (submitted to *Geophysics*).

Mohan NL, Reddy KG and seshagiri Rao SV. 1991. Use of scale and shift invariances of the Mellin-Fourier transform in seismics. *Expanded Abstracts, 61st Intl. Ann. Mtg Soc. Exploration Geophysicists*, pp. 1359-1363.

Scholz CH and Mandelbrot BB. 1989. *Fractals in Geophysics*. Birkhäuser-Verlag, Berlin.

Schuster HG. 1988. *Deterministic Chaos: An Introduction*. VCH Verlagsgesellscraft, Weinheim, Germany.

Stevens RT. 1993. *Advanced Fractal Programming in 'C'*. BPB Publ., N. Delhi, India.

APPENDIX A

Scale-Translation Group 1-D—Subgroup of Scale-Euclidean Group

$$t' = t - a \quad \rightarrow \quad \text{translation}$$
$$t' = e^\lambda t \quad \rightarrow \quad \text{scale transformation or dilation}$$

Scaling - A Rotation

Here, by keeping the origin fixed, the $X - Z$ co-ordinate axes are denoted as $X' - Z'$. Let (x, z) and (x', z') be co-ordinates of a point P in $X - Z$ and $X' - Z'$ planes respectively. Then the relations connecting (x, z); (x',z') and O (Fig. 1) are given by (Mohan *et al.*, 1991)

$$x' = x \cos \theta + z \sin \theta$$

and
$$z' = -x \sin \theta + z \cos \theta \tag{A-1}$$

If the origin is shifted to the point (m, n) and rotated through an angle θ, then the transformation formulae are given by

$$x' = (x - m) \cos \theta + (z - n) \sin \theta$$

and
$$z' = (z - n) \sin \theta - (x - m) \cos \theta \tag{A-2}$$

APPENDIX B

Fourier Transform

$$F(w) = \int_{-\infty}^{\infty} f(t) e^{-iwt} \, dt \tag{B-1}$$

$$f(t) = \frac{1}{2\pi} \int_{-\infty}^{\infty} F(w) e^{iwt} \, dw \tag{B-2}$$

Mellin Transform

$$M(s) = \int_{0}^{\infty} f(t) t^{s-1} \, dt \tag{B-3}$$

$$f(t) = \frac{1}{2\pi i} \int_{c-i\infty}^{c+i\infty} M(s) t^{-s} \, ds \tag{B-4}$$

Mellin Transform in the Complex Domain

$$M(is) = M(s) = \int_{0}^{\infty} f(t) \, t^{-is-1} \, dt \tag{B-5}$$

$$f(t) = \frac{1}{2\pi i} \int_{c-i\infty}^{c+i\infty} M(s) t^{-is} \, ds \qquad \text{(B-6)}$$

Mellin-Fourier Transform

$$M(s) = k^{-is} \int_{-\infty}^{\infty} f(ke^{p}) \, e^{-ips} \, dp; \qquad [k^{-is} = 1] \qquad \text{(B-7)}$$

MT of exponentially distorted input function is converted as M-F transform.

APPENDIX C

Transform Properties

Shift Invariance

$$\begin{array}{cc} f_1(t) & f_2(t - t_0) \\ \downarrow & \downarrow \\ F_1(w) & e^{-iwt_0} F_2(w) \end{array}$$

$$|F_1(w)| = |F_2(w)|$$

Scale Invariance

$$\begin{array}{cc} f_1(t) & f_2(at) \\ \downarrow & \downarrow \\ M_1(s) & a^{-s} M_2(s) \\ \downarrow & \downarrow \\ M_1(s) & e^{-i\lambda s} M_2(s) \end{array} \qquad a = e^{i\lambda}$$

$$|M_1(s)| = |M_2(s)|$$

Shift and Scale Invariance

$$\begin{array}{cc} f_1(t) & f_2[a(t - t_0)] \\ \downarrow & \downarrow \\ F_1(w) & e^{-iwt_0} F_2(w/a) \end{array}$$

(considering only magnitude of the Fourier transform)

$$\begin{array}{cc} M_1(s) & e^{-i\lambda s} M_2(s) \end{array}$$

$$|M_1(s)| = |M_2(s)|$$

APPENDIX D

Mellin-Power Spectrum

Analogous to the Fourier power spectrum, the Mellin power spectrum contributes to the input function $f(t)$. The spectral contribution at $s = s_1$ to $f(t)$, is given by (Mohan *et al.*, 1991),

$$f(t)|_{s = s_1} = |(1/\pi) A \ |t| \ \exp(-v\pi s/2).$$

$$\cos \ (s \log \ |t| + v\pi/2 + \phi)|_{s = s_1} \qquad \text{(D-1)}$$

where

$$n = \text{sgnt}, \qquad A = |M(s) \overline{|(-is)}|_{s = s_1}$$

and ϕ is given by $M(s) \overline{|(-is)} = A \exp(-i\phi)$

The property of $f(t)|_{s = s_1}$ is given by

$$|\exp(2\pi n/s) \ f(\exp((-2\pi n/s)t) \ |_{s = s_1} = f(t) \ |_{s = s_1'} \qquad \text{(D-2)}$$

$$(\pm \ n \text{ integer })$$

A peak in the Mellin power spectrum contributes to $f(t)$. By multiplying $f(t)$ with the scale factor $\exp(2n\pi/s)$, we obtain the original feature, where the feature is stretched for $+n$ and compressed for $-n$. Also, the stretching and compressing facilitate to realise the perodicity of the original feature.

APPENDIX E

M-F Transform of Analytic Wavelet (Scaled)
    The generalised analytic Rayleigh wavelet is given by (Hubral and Tygel, 1989),

$$_N B_n^{f_M \ d} (t) = (-1)^n \ \exp(i \ (\alpha + \pi/2 \ ))/(i + (2\pi f_M \ t/n) \ )^{n + 1} \qquad \text{(E-1)}$$

$$t \to \text{time}; \quad f_M \to \text{main frequency}$$

$$n \to \text{order of wavelet } (\pm \text{ve integer})$$

and phase angle $\alpha = (2\pi/360°)r$; $r = 0°, 10°, 20°, ..., 170°$. By simple change in polarity, the wavelet can be constructed for $r > 170°$. The Mellin-Fourier transform of the scaled analytic wavelet is given by (Mohan *et al.*, 1991),

$$M(s) = 2\pi \ i^n \overline{|(s + n)} \exp(-i(\phi s + \alpha + \pi/2)) \ \varepsilon^{-s}/\overline{|(n + 1)} \qquad \text{(E-2)}$$

The magnitude

$$|M(s)| = \sqrt{RI^2 + IM^2}$$

$$= 2\pi \overline{|(s + 1)} \ \varepsilon^{-s} \qquad (n = 1) \qquad \text{(E-3)}$$

and the phase

$$P(s) = \arctan\left[\tan\left\{(\phi s + \alpha + \pi/2) + \pi/2\right\}\right]$$

$$= \pi + \alpha + \phi s \qquad \text{(E-4)}$$

APPENDIX F

Phase component of the M-F Transformation of Seismic Wave Form— Analyticity

The phase component of the M-F transform of the seismic wave signature is given by

$$P(s) = \arctan[-\mathrm{Im}M(s)/\mathrm{Re}M(s)\,] = -s(p + iq) \qquad \text{(F-1)}$$

where $p = \ln\left(\sqrt{x^2 + iz^2}\right)$ and $q = \arctan(-z'/x')$

Differentiating partially the real and imaginary components (eqn (F-1)) with respect to x' and z', we write

$$\partial p/\partial x' = sx'/(x'^2 + z'^2); \qquad \partial p/\partial z' = sz'/(x'^2 + z'^2) \qquad \text{(F-2)}$$

and
$$\partial q/\partial x' = sz'/(x'^2 + z'^2); \qquad \partial q/\partial z' = -sx'/(x'^2 + z'^2)$$

From eqn (F-1), it is evident that

$$\partial p/\partial x' = -\partial q/\partial z'; \qquad \text{and} \qquad \partial p/\partial z' = \partial q/\partial x' \qquad \text{(F-3)}$$

and therefore satisfy the Cauchy-Reimann conditions throughout and hence $P(s)$ is analytic in the domain D. If $P(s)$ is analytic in a domain D, its components p and q are harmonic in D.

We define the $P(s)$ as the logarithmic function of complex analysis by the equation

$$\log Z = \log r + i\theta \qquad \text{(F-4)}$$

CAN TRAVEL-TIME CURVE TUNNEL THROUGH CHAOTIC REGIME

Multiple reflections is a good subject for nuclear physicists, astrophysicists, and mathematicians who enter our field. Those who are willing to take up the challenge of trying to carry theory to industrial practice are rewarded by some humility.

— J.F. Claerbout

Dedicated to J.F. Claerbout.

N.L. Mohan* and L. Anand Babu**

INTRODUCTION

The present intuitive concept of travel-time ray paths is based on certain important features of wave propagation through a complex geological regime. We hypothesise that complex geological settings are chaotic and fractalised (Figs. 1 and 2) (Fig. 2, Ferril and Morris, 1997; Fig. 6, Tari *et al.*, 1997; Mohan and Anand Babu, 1997a, b) from the structural and compositional point of view (see Appendix for important terminologies pertaining to chaos and fractals). The argument (Korvin, 1992) is that (i) waves interacting with fractals become fractals themselves; (ii) waves that have encountered fractals were termed diffractals; (iii) the concept constitutes a new regime in wave propagation theory, because conventional perturbation and geometrical optics do not apply; (iv) waves acquire random

* Centre of Exploration Geophysics, Osmania University, Hyderabad 500 007, India.
** Department of Mathematics, Osmania University, Hyderabad 500 007, India.

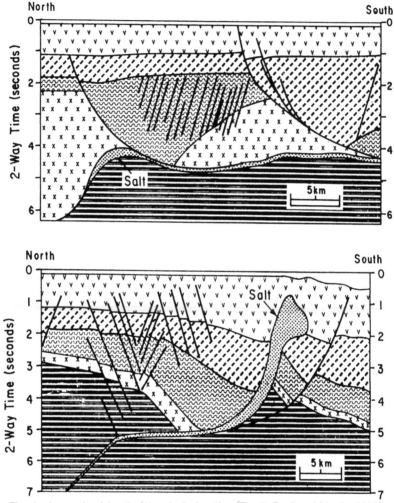

Fig. 1 A complex (chaotic ?) geological setting [Fig. 2; Ferril and Morris, 1997.]

fluctuations in their wave functions ϕ and intensities $I = |\phi|^2$ when they encounter random structures S; (v) self-similar objects look the same when space is stretched (dilated) uniformly by a factor λ, and (vi) self-affine objects look the same after the affine transformation $X \Rightarrow (x_1, x_2, x_3) \Rightarrow X' \Rightarrow (r_1 x_1, r_2 x_2, r_3 x_3)$ where the scaling ratios r_1, r_2, r_3 are not at all equal.

Further, based on Marmousi data set (Figs. 3 and 4) (Figs. 2 and 3, Geoltrain and Brac, 1993), we mention that the pathologies of the travel-time field reveals that the first arrival (Fig. 5(a)) is often a marginally energetic event whenever subsequent arrival occurs. Multiple arrival occurs whenever the geological structure creates strong and localised

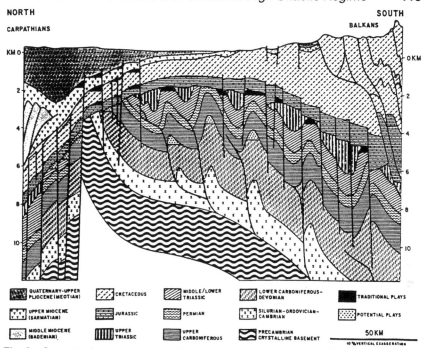

Fig. 2   Conceptual complex (chaotic ?) geological section across the Moesian platform of Romania and Bulgaria [Fig. 6; Tari *et al.*, 1997]

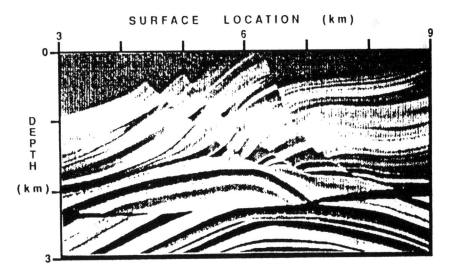

Fig. 3   Marmousi data set: velocity field showing structure [Fig. 2; Geoltrain and Brac, 1993]

velocity heterogeneities (Fig. 5(b)). Therefore it is incorrect to visualise the travel-time as single valued; in fact it associates two major properties, namely irregularity and incompleteness.

Irregularity occurs (a) in the form of discontinuities in travel-time gradients (i.e., propagation direction; Fig. 6) (Fig. 4, Geoltrain and Brac, 1993), (b) in the form two arrivals have equal travel-time even after two different paths and (c) with respect to source position at a fixed depth location generates 'migration smiles' (Figs. 7 to 9) (Figs. 17, 18, and 20; Geoltrain and Brac, 1993).

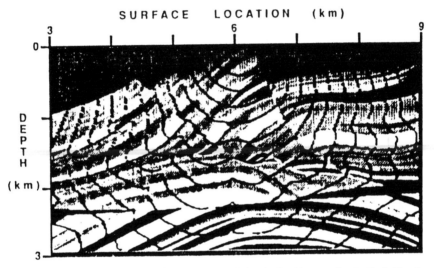

Fig. 4   First arrival wavefronts for a source at 6 km in the Marmousi velocity field using Podvin's algorithm (time increment between wavefronts is 100 ms) [Fig. 3; Geoltrain and Brac, 1993]

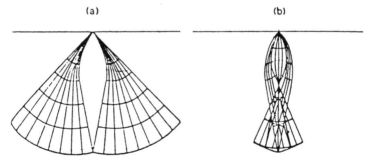

Fig. 5   Ray families associated with: (a) the first arrival ; (b) later arrivals [Fig. 16; Geoltrain and Brac, 1993]

Incompleteness occurs (i) when there are multiple arrivals (Fig. 5(b)), (ii) due to mispositioning at depth with respect to faster first arrival and stacking of prestack migrated sections yields an incoherent image.

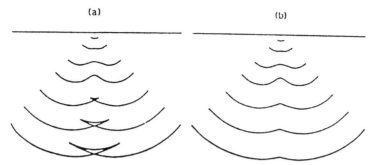

Fig. 6   A simple example of multivalued travel-time field: (a) complete triplicated wavefront; travel-time gradient is continuous along the front; (b) first arrival wavefront; travel-time gradient is discontinuous and only the fastest front is computed [Fig. 4; Geoltrain and Brac, 1993]

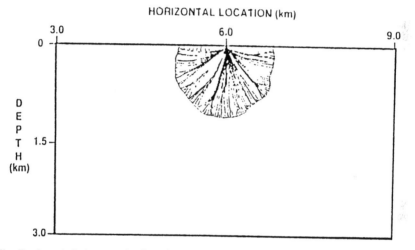

Fig. 7   A posteriori ray tracing from isochron 500 ms (source at 6 km) [Fig. 17; Geoltrain and Brac, 1993]

The present hypothesis is that the internal layers of the earth media are geometrically irregular and heterogeneous (from the point of velocity, density etc.) and as a result the assumption of linear model is no longer valid. Entertaining a non -linear model depends on several factors, such as source energy, depth, geometry(or shape) of the geological target, vertical/lateral velocities, incident/reflected (or refracted/diffracted) angles, surface topography, geometrical set-up of source-receiver configuration etc. Scattering of energy might further complicate the evaluation of parameters such as velocities, shape of wave signatures, arrival times, depth etc. (Mereu and Ojo, 1981; Crossly and Jensen, 1989; Korvin, 1992; Mohan *et al.*, 1991; Mohan and Reddy, 1991; Jin and Madariaga, 1994; Jervis *et al.*, 1996).

HORIZONTAL LOCATION (km)

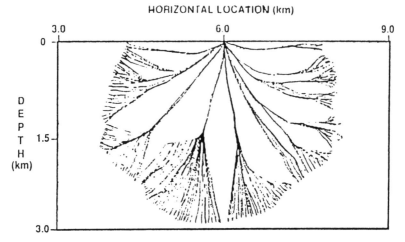

Fig. 8 A posteriori ray tracing from isochron 1000 ms (source at 6 km) [Fig. 18; Geoltrain and Brac, 1993]

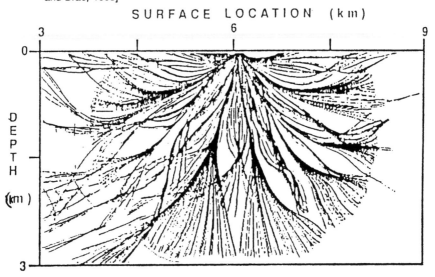

Fig. 9 First-arrival and source ray families of Marmousi velocity field [Fig. 20; Geoltrain and Brac, 1993]

Here we attempt to construct the travel-time paths, essentially based on the sensitive dependence on initial time impulse function (depth/velocity), without assuming the depth of layers and the corresponding velocities.

LOGISTIC TRAVEL-TIME MAP

Verhulst in 1845 (Schuster, 1988) formulated the growth of population in a closed area as

$$x_{n+1} = f_r(x_n) = r\, x_n\, (1 - x_n) \tag{1}$$

i.e. as the population number increases (x_{n+1}) the area gets diminished with respect to the previous number (x_n). The parameter r depends on fertility, the actual area of living etc. Grossmann and Thomae (1977), Feigenbaum (1978) and others found that the iterates x_1, x_2,... of eqn. (1) display as a function of external parameter r, a rather complicated behaviour that becomes 'chaotic' at large r.

Following eqn. (1) above we reconstruct

$$(z/v)_{n+1} = f_r[(z/v)_n] = r(z/v)_n\, (1 - (z/v)_n) \tag{2}$$

where $(z/v)_0$ is time-impulse function. A note is in order that the travel-time path in complex geological regime is equally complex and irregular. We therefore believe that the external parameter, r, is also complex and it may have amplitude and phase components. Accordingly, considering r as a complex number, we recast eqn. (2) as

$$(z/v)_{n+1} = f_r[(z/v)_n] = (r_1 + ir_2)(z/v)_n\, (1 - (z/v)_n) \tag{3}$$

Now we define the amplitude as

$$AM((z/v)_{n+1}) = \sqrt{[\mathrm{Re}\,((z/v)_{n+1})]^2 + [\mathrm{Im}\,(z/v)_{n+1}]^2} \tag{4}$$

and phase

$$PH((z/v)_{n+1}) = \mathrm{arc\ tan}\,[\mathrm{Im}((z/v)_{n+1})/\mathrm{Re}((z/v)_{n+1})] \tag{5}$$

At this stage it is important to understand the concept—how to characterise chaotic motion. Here we introduce the Lyapunov exponent, which is generated by one-dimensional Poincare maps, as a quantitative measure to characterise chaotic motion.

LYAPUNOV EXPONENT

In the iterative process of any physical system (dynamical system) we find a certain amount of deviation or separation from one iteration to another and the amount of separations depends on the initial parameters or conditions. These separations lead to chaotic motion. If the separation is exponential in nature, then such separation is called the Lyapunov exponent (Schuster, 1988) (Fig. 10).

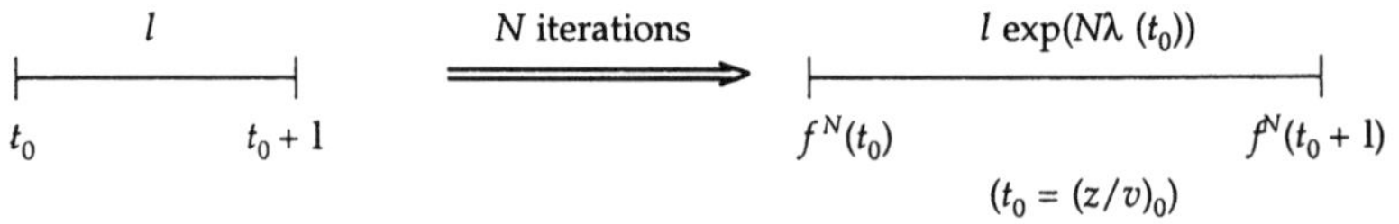

Fig. 10 Concept of Lyapunov exponent

Fig. 10 gives us

$$l \exp(N\lambda(t_0)) = |f^N (t_0 + 1) - f^N(t_0)|$$

i.e.,

$$\lambda(t_0) = \lim_{N\to\infty} \lim_{l\to 0} (1/N) \log |[f^N(t_0 + 1) - f^N(t_0)]| / l$$

$$= \lim_{N\to\infty} (1/N) \log |df^N (t_0)/dt_0| \tag{6}$$

It indicates that $\exp(\lambda(t_0))$ is the average factor by which the distance between closely adjacent points becomes stretched after one iteration. Further, the chain rule of eqn. (6)

$$df^2(t)/dt \,|_{t=0} = df[f(t)]/dt \,|_{t=0} = f'[f(t_0)] f'(t_0)$$

$$= f'(t_1) f'(t_0); \qquad t_1 \equiv f(t_0)$$

Now we write the Lyapunov exponent as

$$\lambda(t_0) = \lim_{N\to\infty} (1/N) \log |df^N(t_0)/dt_0|$$

$$= \lim_{N\to\infty} (1/N) \log \left| \prod_{i=0}^{N-1} |f'(t_i)| \right.$$

$$= \lim_{N\to\infty} (1/N) \sum_{i=0}^{N-1} \log |f'(t_i)| \tag{7}$$

Loss of information: We consider the interval $[0, 1]$ in n equal intervals and assume that point t can occur in each of them with equal probability $1/n$. By guessing which interval contains t , we gain the information

$$I_0 = -\sum_{i=1}^{n} (1/n) \log_2 (1/n) = \log_2 (1/n) \tag{8}$$

I is reduced if we decrease n, and $I = 0$ for $n = 1$. As an example we consider a linear map $f(t)$ that changes the length of an interval by a factor $a = |f'(0)|$. The corresponding decrease of resolution leads to a loss of information after the mapping, given by:

$$\Delta I = -\sum_{i=1}^{n/a} (a/n) \log_2 (a/n) + \sum_{i=1}^{n} (1/n) \log_2 (1/n) = -\log_2 a$$

$$= -\log_2 |f'(0)|$$

Similarly we generalise the above expression to a situation wherein $|f'(t)|$ varies from point to point and averaging over many iterations leads to the following for the mean loss of information

$$\overline{\Delta I} = -\lim_{N\to\infty} (1/N) \sum_{i=0}^{N-1} \log_2 |f'(t_i)| \tag{9}$$

and is proportional to the Lyapunov exponent,

$$\lambda(t_0) = (\log 2) \cdot |\overline{\Delta I}| \tag{10}$$

This relation between the Lyapunov exponent and the loss of information is a first step towards characterisation of chaos in a co-ordinate invariant system (Schuster, 1988).

MODEL SIMULATION

We construct the models with the following set of parameters:

(1) $z = 50$ m ; $v = 1500$ m/s; number of iterations = 1500; ($2.8 < r < 4$);

(2) $z = 75$ m ; $v = 1800$ m/s; number of iterations = 2000; $r_1 = 1$ (0.1) 2.5; $r_2 = 2$ (0.1) 3.5 ; and

(3) $z = 75$ m; $v = 2000$ m/s; $\theta = 30^0$; $1 < r < 4$; number of iterations = 4000.

We mention here that the iterative process, using the scale parameter, is similar to 'the source repeatedly fired at 100 m/s intervals to generate expanding wavefronts in the medium' (Geoltrain and Brac, 1993).

We show the plots of logistic maps (1) $(z/v)_{n+1}$ versus r; (2) $AM((z/v)_{n+1})$ versus $|r|$, $Re((z/v_{n+1}))$ versus r_1, $Im((z/v)_{n+1})$ versus r_2, $PH((z/v)_{n+1})$ versus $|r|$, and Lyapunov exponent $\lambda((z/v)_0)$ versus $|r|$; and (3) $((z/v \cos (\theta))_{n+1})$ versus r , Lyapunov exponent $\lambda((z/v \cos (\theta)_0))$ versus r, and loss of information $(\overline{\Delta I})$ versus r for the corresponding models in Figs. 11; 12(a)-(e); and 13 (a)-(c).

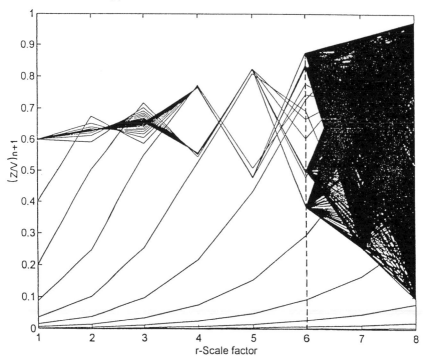

Fig. 11 Logistic travel-time ray paths $[((z/v) \times e)_{n+1}]$ versus scale factor r

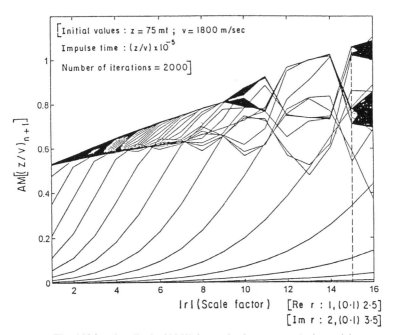

Fig. 12(a) Amplitude AM $[(\langle z/v \rangle \times e)_{n+1}]$ versus scale factor $|r|$

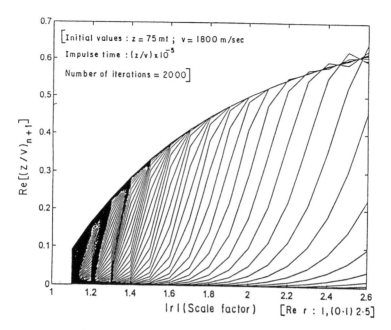

Fig. 12(b)   Real component Re $[((z/v) \times e)_{n+1}]$ versus scale factor r

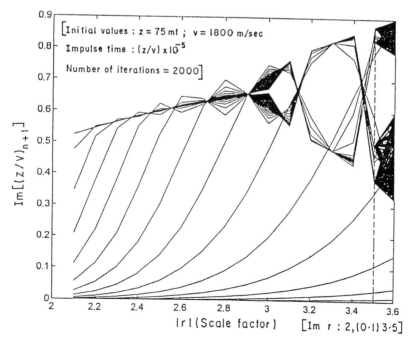

Fig. 12(c)   Imaginary componet Im$[((z/v) \times e)_{n+1}]$ versus scale factor r

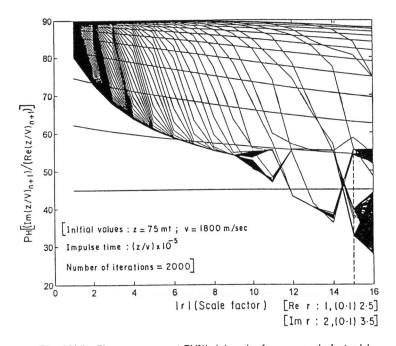

Fig. 12(d) Phase component PH[((z/v) × e)$_{n+1}$] versus scale factor |r|

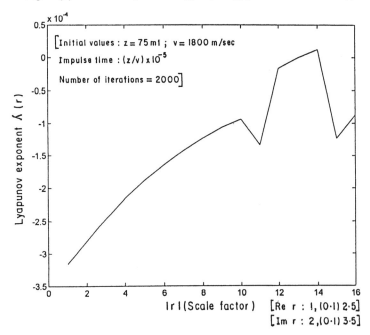

Fig. 12(e) Lyapunov exponent λ(t_0) versus scale factor |r|; [t_0 = ((z/v) × e)$_0$]

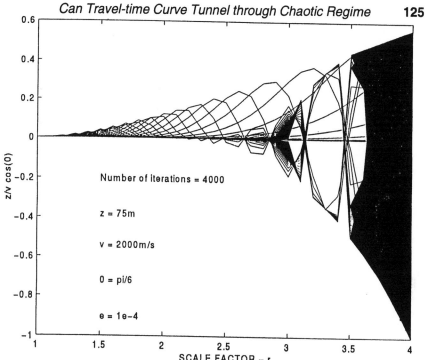

Fig. 13(a) Logistic travel-time ray paths $[(z/v \cos \theta \times e)_{n+1}]$ versus scale factor r

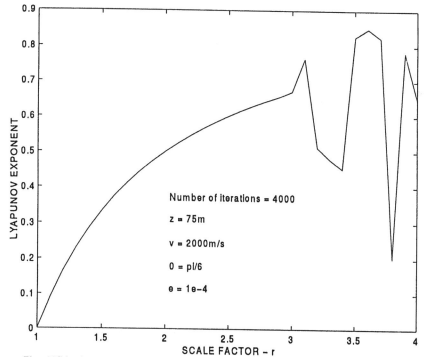

Fig. 13(b) Lyapunov exponent $\lambda(t_0)$ versus scale factor r; $(t_0 = [((z/v) \cos \theta) \times e)_0])$

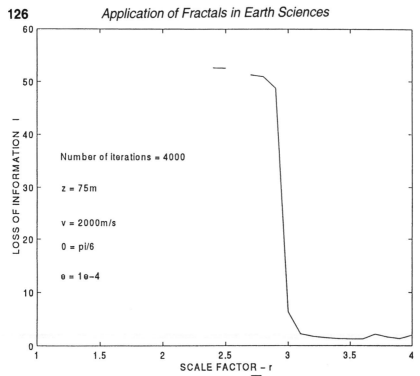

Fig. 13(c)   Mean loss of information $(\overline{\Delta I})$ versus scale factor r

ANALYSIS

From Figs. 11; 12(a), 12(c), 12(d); 13(a) and 13(c) we observe two distinct features: (1) 'bifurcation or periodic regime' for $1 < r < r_\infty$ and (2) 'chaotic regime' for $r_\infty < r \le 4$. These regimes are characterised as follows (Schuster, 1988):

(a) Periodic regime

1. The scaling of values r_n (from one bifurcation to another bifurcation) takes place as the number of fixed points changes from 2^{n-1} to 2^n as

$$r_n = r_\infty - \text{const. } \delta^{-n}, \qquad \text{for } n \gg 1 \qquad (11(a))$$

2. The distances d_n (i.e., the vertical distances within the bifurcations with respect to a fixed point) of the point in a 2^n-cycle that are closest to $t = 1/2$ (Fig. 14) have constant ratios

$$-\alpha = d_n / d_{n+1}, \qquad \text{for } n \gg 1 \qquad (11(b))$$

(related to the scaling of the distance between the iterates)

That is, we understand that simultaneous rescaling takes place along the X- and Y-axis and is possible only with the doubling transformation $[Tf(t) = -\alpha f [f(-t/\alpha)]$.

We denote these two values δ and α as Fiegenbaum constants, which are universal, and have

$$\delta = 4.66692016091... \quad \text{and} \quad \alpha = 2.5029078750... \qquad (11(c))$$

Further, we also observe (Fig. 14) the distances between the vertical scaling (with respect to a fixed point) along the $|r|$ axis, similar to r_n, is R_n and we define

$$R_n - r_\infty = \text{const. } \delta^{-n}; \quad \text{and} \quad R_\infty = r_\infty = 3.5699456... \qquad (11(d))$$

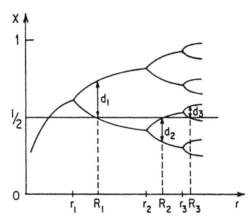

Fig. 14 Distances d_n of the fixed points closest to $x = 1/2$ for superstable 2^n-cycles (schematically) (Fig. 24, Schuster, 1988).

(b) Chaotic regime

1. The chaotic intervals move together by inverse bifurcation until the iterates become distributed over the interval [0, 1] at $r = 4$.
2. The periodic p cycles ($p = 3, 5, 6...$), a characteristic feature of r windows, with successive bifurcations $p, p \cdot 2^1, p \cdot 2^2$ etc. The corresponding r-values scale like (eqn. 11(a)) with the same but different constants.
3. Different Fiegenbaum constants, which are also universal, occur for period triplings $p \cdot 3^n$ and quadruplings $p \cdot 4^n$ etc. as $\bar{r} = \bar{r}_\infty - \text{const. } \delta^{-n}$ (e.g. $\delta = 55.247...$ for $p \cdot 3^n$).

HENON TRAVEL-TIME MAP

While dealing with multivalued dip move out (DMO) analysis, Artley and Hale (1994) did not make assumptions about the offset, dip, velocity function or hyperbolic move-out. Since the velocity of the medium varies with depth only, ray tracing (Fig. 15) need be performed for only a single surface location and rays do not need to be traced from every surface

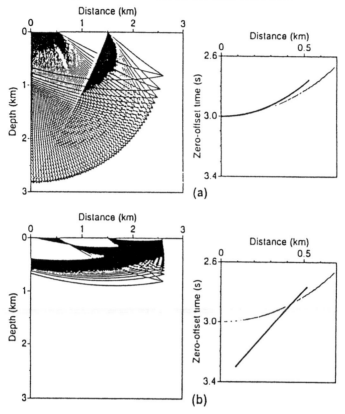

Fig. 15 Physical interpretation of the multivalued DMO operator: ray trios corresponding to the (a) primary, (b) secondary branch operators [Fig. 9; Artley and Hale, 1994]

location. Accordingly, Artley and Hale (1994) obtained the solution by framing a system of five non-linear equations with six unknowns. Against this background, we treat the problem as a ray tracing through a chaotic geological regime.

Henon (1976) defined the two-dimensional analogue of the logistic map as (Schuster, 1988)

$$x_{n+1} = 1 - a\,x_n^2 + y_n \qquad \text{or} \qquad x_{n+1} = 1 - ax_n^2 + bx_{n-1} \qquad (12)$$

$$y_{n+1} = bx_n$$

where a and $|b| < 1$ are external parameters.

In order to simulate travel-time curves we formulate the above given Henon map equation as

$$(z/v)_{n+1} = 1 - a(z/v)_n^2 + b\,(z/v)_{n-1} \qquad (13)$$

where $(z/v)_0$ is time-impulse function.

MODEL SIMULATION

We construct the Henon travel-time ray path models with the following set of parameters;
(1) $z = 20$ m; $v = 2000$ m/s; number of iterations = 1000; $e = 10^{-5}$; initial time-impulse value = $(20/2000) \times e$, and
(2) $z = 50$ m; $v = 2500$ m/s; number of iterations = 1000; $e = 10^{-5}$; initial time-impulse value = $(50/2500) \times e$.

The travel-time ray paths originate vertically down from the source point. However, because of the expansion and contraction nature of the Henon travel-time field, the ray path rebounds to the surface away from the point of source(i.e., along the line from the source point on the surface). Here, the triangulation of ray path trajectories, overlapping each other, form a cone, a sort of polarisation (Figs. 16 and 17).

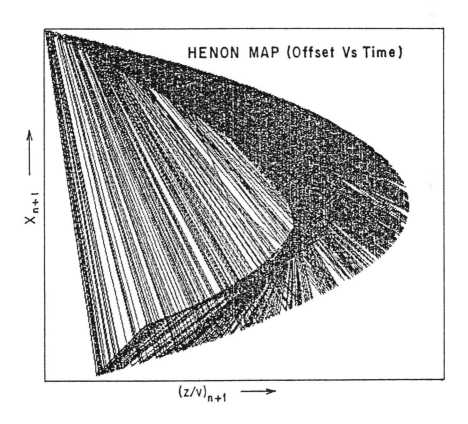

Fig. 16   Henon travel-time ray paths, model 1

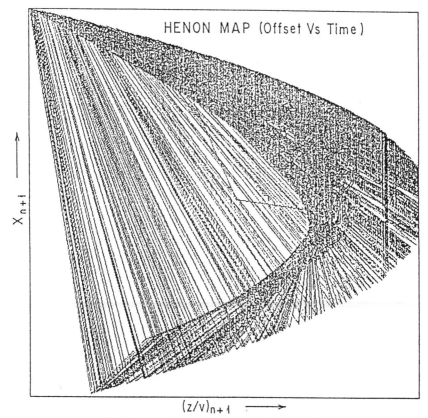

Fig. 17   Henon travel-time ray paths, model 2

ANALYSIS

The computed Henon travel-time map (Figs. 16 and 17) is area contracting, i.e., is dissipative for $|b| < 1$ because its Jacobian is just

$$\det\left|\begin{pmatrix} -2ax_n & 1 \\ b & 0 \end{pmatrix}\right| = |b| \tag{14}$$

For instance decomposition of the action of a Henon map on an ellipse illustrates (Fig. 18) (Fig. 71, Schuster, 1988) how area bending, contraction etc. take place.

Further, the Henon travel-time map is (a) chaotic, (b) attracted to a bounded region in phase space, (c) inhomogeneous, (d) contracts in one direction and stretches in other direction, and (e) self-similar. Maps possessing the said characteristics are called 'Henon Attractor' (also, see Fig. 72; Schuster (1988) for self-similar structure of Henon attractor).

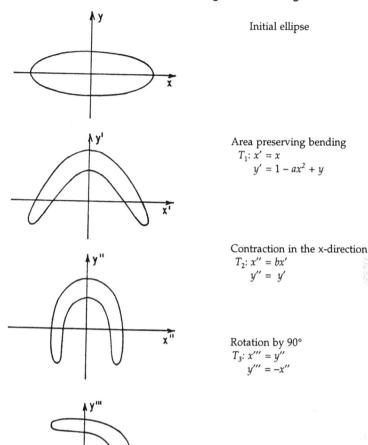

Initial ellipse

Area preserving bending
T_1: $x' = x$
$\quad\;\; y' = 1 - ax^2 + y$

Contraction in the x-direction
T_2: $x'' = bx'$
$\quad\;\; y'' = y'$

Rotation by 90°
T_3: $x''' = y''$
$\quad\;\; y''' = -x''$

Fig. 18 Decomposition of the action of a Henon map $T = T_3 \cdot T_2 \cdot T_1$ on an ellipse (Fig. 71; Schuster, 1988)

The Henon travel-time path spreads through chaotic geological strata due to either structurally highly disturbed areas or random velocity variations or both. In fact, the present simulation (Figs. 16 and 17) substantiates the concept of chaotic travel-time regime when compared with a part of the constructed travel-time paths, spread over a part of offset distance (Fig. 19) (Fig. 10, Artley and Hale, 1994).

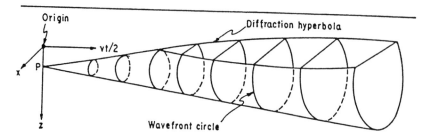

Fig. 19 Point diffractor at *P*. The space-time distribution of energy lies on the cone. The energy observed at the surface *z* = 0 lies on the diffraction hyperbola, whereas at any fixed instant of time the energy lies on a wavefront circle (Robinson, 1983)

DISCUSSION

(A) Logistic travel-time map

From the constructed logistic travel-time ray paths, originated (i) vertically down below the source with real (Fig. 11); and complex scale parameter (Figs. 12(a) – (c)) and (ii) at an angle with respect to the vertical from the source point (Fig. 13(a)), we notice certain interesting features: (a) only up to a certain depth extent travel-time ray paths, irrespective of the value of scale parameter *r*, merge as a single ray, i.e., a line singularity, which indicates that the media are uniform and homogeneous and velocity gradually increases; (b) irregularity in propagation direction of travel-time fields occurs in the form of discontinuities whenever two arrivals have more or less equal travel-time ray paths after following two different paths (Figs. 11, 12(a) and 13(a)); (c) interestingly, the multiples are due to the manifestation of scaled travel-time ray paths; (d) from a few significant converging points of travel-time ray paths (with different values of scale parameter *r*) we expect substantial contrast in velocity field; (e) the converging point of ray paths implies a very strong reflection event; (f) the higher the scale parameter *r* the stronger the energetic events and never faster than the first arrival; (g) from the point where the disappearance of line singularity in the first arrival field occurs, bifurcation in travel-time field starts (i.e., creates a shadow zone) but the travel-time field is more or less equal on either side of bifurcation and we find two diffraction events. At this point of time a question arises: Is it possible and meaningful to record the wave signatures on a negative time axis with respect to the source point? The answer, in our opinion, appears to be 'yes'. Further, we observe that after a few bifurcations, the travel-time field enters the chaotic regime where thick overlapping of all bifurcations of ray paths takes place and what Geoltrain and Brac (1993) have coined as 'migration smiles'. This implies that wave energy totally collapses, no matter the act of scale parameter *r* and number of iterations might be.

Complex scale regime

We notice some more interesting features from the logistic travel-time ray paths, using the complex scale parameter $r_1 + ir_2$. The real components of the ray paths (Fig. 12(b)) show that they are less energetic, irrespective of the value of scale parameter r, and all ray paths merge as a single ray, i.e., a line singularity, and a multiple occurs only at a much later stage. This is due to the narrow scale range $[r_1 = 1.0 \ (0.1) \ 2.5]$. However, the imaginary component of the logistic ray paths (Fig. 12(c)) $[r_2 = 2.0 \ (0.1) \ 3.5]$ are more energetic, the line singularity appears till the value of scale parameter reaches 2.9; subsequently bifurcation begins and ray paths enter a chaotic regime as $r \rightarrow 3.5$.

The amplitude of travel-time field versus $|r|$ (Fig. 12(a)) significantly gives a deeper insight of ray path characters in heterogeneous media. The line singularity, a convergence of ray paths emanated due to different scale values of $|r|$, occurs up to $|r| \approx 2.1$. Later, bifurcation starts with two distinct boundaries, one with the convergence of rays generated with lower scale parameter $|r|$ as an upper boundary, and the lower boundary with the convergence of rays with higher scale value $|r|$. It is an obvious indication that the upper boundary of converged ray paths is less energetic and travels fast, whereas the lower boundary moves slowly, because it associates with multiples and is more energetic.

A stunningly significant point is that in shape the bifurcated parts of ray path boundaries (Fig. 12(a)) are diffracted hyperbola and in fact exactly similar to the intuitive concept (Fig. 19) (Fig. 2.14, Robinson, 1983).

Phase portrait and Lyapunov exponent

It is remarkably evident from the phase portrait of complex travel-time field distribution (Fig. 12(d)) that initially ray path trajectories converge and move vertically down (depth direction) with respect to the source point for a small segment of time. Later, ray paths spread angularly with respect to the vertical. The trajectory of converged ray paths, which form a singularity, bend to the extent of $\approx 50°$ as scale parameter $r \approx 2.2$. Further, a few bifurcations take place ($> 50°$) and finally plunge into the chaotic regime, where 'migration smiles', (Geoltrain and Brac, 1993) (Fig. 12(d)) occur.

The Lyapunov exponent curve (Fig. 13(b)) explicitly indicates the characteristic features of the travel-time curves (or trajectory of ray paths) in complex velocity media, where the chaotic regime (or 'migration smiles') begins as $r \rightarrow 3.5$. Further, the mean loss of information $(\overline{\Delta I})$ versus scale parameter r, confirms the behaviour of travel-time field in complex media of the earth.

It is evident from the constructed logistic travel-time curve (Figs. 11 to 13) that the multiples, occurring due to the variation of scale parameter r, are exactly similar to the intuitive multiple travel-time curves (Fig. 5(b)) (Fig. 16 , Geoltrain and Brac, 1993). The logistic travel-time paths (Figs. 11 to 13) strongly establish the fact that the Marmousi velocity field falls under the chaotic regime.

Also, we observe from the logistic travel-time ray paths (Fig. 13(a)), generated with an angle 30° with respect to the vertical from the source point with different values of scale parameter r, merge as a singularity for a very short span of time and subsequently form as a sort of meshed envelope. However, we notice a few set of multiples, diffraction events, shadow zones and chaotic regime ($r_\infty > 3.5$). In comparison, with respect to Fig. 11, we see in Fig. 13(a) a distinct feature of travel-time ray paths pushing towards the positive side with respect to the vertical. Also, it is evident from the Lyapunov exponent $\lambda(t_0)$ (Fig. 13(b)), and the mean loss

of information $(\overline{\Delta I})$ (Fig. 13(c)) that the travel-time trajectories can be estimated only up to the universal constant r_∞.

(B) Henon travel-time map

We believe that the repeated bouncing of 'Henon travel-time ray paths' may possibly be due to several factors: (i) narrow geological zone between two oppositely inclined fault planes, (ii) a high velocity geological strip, (iii) anisotropic geological column and (iv) highly fragmented geological zone that facilitates strong scattering of wave energy.

Henon ray path phenomena trigger either because of one or more combinations of complex geologically fragmented structures/velocities. Perhaps, this type of contracting and stretching ray path trajectories (Henon ray path map) may be helpful in tomographic scanning of a complex geological object. At this juncture we believe that there is no physical basis for occurrence of irregular travel-time trajectories but rather a typical pathology of first arrival travel-time fields in the presence of multiple arrivals. That is, it is essentially due to non-linear earth media.

In the case of travel-time paths (Figs. 16 and 17) using the Henon map construction, these reflect the intuitive concept of expansion and contraction (Fig. 18) (Fig. 71, Schuster, 1988). Also, the Henon travel-time paths (Figs. 16 and 17) are exactly similar to the ray trios constructed for interpretation of the multivalued DMO operator (Fig. 15) (Fig.10; Artley and Hale, 1994). Further, the constructed Henon map (Figs. 16 and 17) resembles a part of the travel-time field segment of the ray traces (Fig. 20) (Fig. 10, Artley and Hale, 1994).

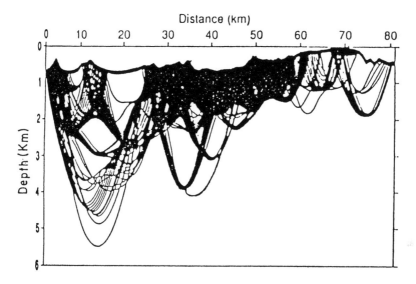

Fig. 20 Some of the 2,500 rays traced through basalt velocity model [Fig. 10; Jarchow *et al.*, 1994].

CONCLUSIONS

The convergence of travel-time paths generated with different values of scale parameter r is bound to yield a significant reflection event irrespective of wave speed, angle of penetration or direction of path. These converged travel-time path events occur randomly as the geometry and geology are heterogeneous. In fact, the constructed ray path trajectories, using logistic and Henon maps correlate well with the Marmousi velocity field (Geoltrain and Brac, 1993) and basalt velocity model (Jarchow *et al.*, 1994) respectively. A primary reflection event might also occur from the point where the travel-time paths bifurcate. In heterogeneous media the paths of wave propagation are different and loss of energy varies and hence so do the amplitudes.

Further, (1) the travel-time path through heterogeneous and fragmented elastic earth media is sensitive to initial conditions and in this context logistic and Henon maps show a singularity (travel for a short span of time), multiples, diffraction events, meshed envelope, shadow zones and chaotic/fractal regimes. And with the complex scale parameter $(r_1 + ir_2)$ components such as amplitude, real, imaginary in general and phase in particular give a deeper insight into the heterogenous media; (2) creation of multiples is essentially due to the act of scale parameter; (3) it might be possible to predict the transition to chaotic regime $(> r_\infty)$; (4) the fractal dimension of the fragmented embedded earth surface (or very fine structural details) can be estimated, as the relation between the Lyapunov exponent $\lambda(t_0)$ and loss of information $(\overline{\Delta I})$ is a first step towards the characterisation of chaos in a co-ordinate invariant system; (5) in a similar way fractal velocity can also be determined; and (6) the travel-time concept may facilitate analysis of seismic data of anisotropic and tomographic studies. Further, this study strongly confirms the Korvin's (1992) arguments (as mentioned in the introduction) pertaining to the travel-time ray path through chaotic and fractal earth media. Further detailed and critical examination of this study may unravel the chaotic and fractalised internal earth structure.

ACKNOWLEDGEMENTS

We profusely thank the Department of Science and Technology, Govt of India, for having sanctioned the financial grant for a research project.

REFERENCES

Artley C and Hale D. 1994. Dip-moveout processing for depth-variable velocity. *Geophysics*, **59(4)**: 610-622.

Crossly D and Jensen OG. 1989. Fractal velocity models in refraction seismology. *PAGEOPH*, **131**: 61-76.

Feigenbaum MJ. 1978. Quantitative universality for a class of nonlinear transformations. *J. Stat. Phys.* **19**: 25.

Ferril DA and Morris AP. 1997. Geometric considerations of deformation above curved normal faults and salt evacuation surfaces. *The Leading Edge*, **16(8)**: 1129-1133.

Geoltrain S and Brac J. 1993. Can we image complex structures with first-arrival travel-time ? *Geophysics*, **58(4)**: 564-575.

Grossmann S and Thomae S. 1977. Invariant distributions and stationary correlation functions of one dimensional discrete process. *Z. Naturforsch*, **32A**: 1353.

Henon M. 1976. A two-dimensional map with a strange attractor. *Commun. Math. Phys.*, **50**: 69.

Jin S and Madariaga R. 1994. Nonlinear velocity inversion by a two-step Monte Carlo method. *Geophysics*, **56**: 577-590.

Jervis M, Sen MK and Stoffa PL. 1996. Prestack migration velocity estimation using nonlinear methods: *Geophysics*, **60**: 138-150.

Korvin G. 1992. *Fractal Models in the Earth Sciences*. Elsevier, NY.

Mereu RF and Ojo SB. 1981. The scattering of seismic waves through a crust and upper mantle with random and vertical inhomogeneities. *Phys. Earth and Planet. Int.* **26**: 233-240.

Mohan NL. 1995. Discussion on 'Pulse distortion in depth migration', *Geophysics*, **60(5)**: 1948-1949.

Mohan NL. 1997. Discussion on 'Regularizing of 3-D data sets with DMO'. *Geophysics*, **62(5)**: 1331-1332.

Mohan NL and Reddy KG. 1991. Application of the Mellin-Fourier transform for pattern recognition of seismic signals. *Expanded Abstracts, 61st Intl. Ann. Mtg. Soc. Exploration Geophysicists*, pp. 1354-1358.

Mohan NL Anand Babu L. 1997a. Can travel-time curve tunnel through chaotic regime. Paper presented at DST Workshop on "Application of Fractals in Geophysics". NGRI, Hyderabad, India.

Mohan NL Anand Babu L. 1997b. Fractality of seismic wave signature—a Mandelbrot approach. Paper presented at DST workshop on "Application of Fractals in Geophysics". NGRI, Hyderabad, India.

Mohan NL, Reddy KG and Seshagiri Rao SV. 1991. Use of scale and shift invariances of the Mellin-Fourier transform in seismics. *Expanded Abstracts, 61st Intl. Ann. Mtg. Soc. Exploration Geophysicists*, pp. 1359-1363.

Robinson EA. 1983. *Migration of Geophysical Data*. D. Reidel Publ. Com., Dordrecht.

Schuster HG. 1988. *Deterministic Chaos*. VCH Publ., NY.

Tari G, Georgiev G, Hardy S, Poblet J and Stefanscu M. 1997. Late triassic Cimmerian structures beneath the miocene platform (Romania/Bulgaria): *The Leading Edge*, **16(8)**: 1153-1157.

APPENDIX

Definitions—Chaos and Fractal

Chaos: Lack of structure and as the manifestation of randomness (in mathematical terminology of dynamical system)—a deterministic phenomenon exhibited by iterated systems and characterised by the effects of nonlinearity.

Fractal/Geometry: A fragmented segment is considered a fractal. The concept. 'fractal' is conceived from self-similarity idea. Fractal geometry— the invariance of the observation with scale is the key to the concept of the role of earth dynamics in the formation and development of hydrocarbon basins.

Dynamic Process: The study of dynamic processes involved in the deformation and fragmentation of the earth with the production of uplifts, downwarps, grabens, block and rift faults etc. The regular appearance of repetitive shapes observable over very wide range of scales—an analysable event characterising the chaotic tectonic field.

Scaling: The essential property of self-similarity is scale invariance, i.e., features observed at one scale can be readily viewed at other scales.

Fractal Geometry-Chaotic Dynamics: The connection between fractal geometry and dynamics concerns the stability of dynamical systems. Dynamical systems are usually described through a set of differential equations and their solutions are generally via iterations. Chaotic observations of late are considered purely deterministic and merely reflect the iterative character of chaos.

Another critical connection between chaos and fractals is uncovered when the chaotic behaviour of a dynamical system is analysed through the use of phase portraits, i.e., displays the projection of orbit in particular phase spaces. For instance, the space using subsequent samples as space co-ordinates.

Poincare's section can be obtained which reveals clusters of points that are intersections of the orbit with the plane selected for the section. These clusters are fractal objects and their fractal dimensions are an important measure of the properties of the dynamic process. Being fractals, these clusters are invariant with scale and as a result determination of significant parameters can be done at one scale and extended to another.

Time Series Analysis: The fragmentation and clustering aspects of fractal objects are intimately connected to the intermittency and persistency of non-stationary time series.

APPLICATON OF FRACTALS IN SEISMOLOGY WITH REFERENCE TO KOYNA EARTHQUAKES

V.P. Dimri*

INTRODUCTION

It has been well demonstrated that the temporal, spatial and magnitude distribution of earthquake populations are scale invariant on some levels and scales, and are fractals in the statistical sense (Kagan and Knopoff, 1980; Kagan, 1981; Aki, 1981; King, 1983). In particular, the magnitude-frequency relation (Gutenberg and Richter, 1954), which is probably the most fundamental relation in earthquake seismology has been restated in terms of fractals (Aki, 1981; King, 1983; Turcotte, 1986). The Gutenberg-Richter's b-value was interpreted in terms of the fractal dimension. A linear correlation has also been observed empirically between the b-value and the correlation dimension of earthquake epicentre distribution (Guo and Ogata, 1997; Oncel et al., 1996). The correlation dimension is a measure of the clustering of an epicentre, while the b-value is related to the power law distribution of earthquake size. It is therefore physically uncertain how these two parameters are related. Any acceptable model for the origin of the b-value should reflect both the dynamic and the geometric properties of faulting. Sornette et al. (1991) and Lomnitz-Adler (1992)

* National Geophysical Research Institute, Hyderabad 500 007, India.

have shown that the b-value can be expressed as a linear combination of the scaling exponent due to the self-organised dynamics of the crust and the fractal dimension of the fault system. This relation, as well as all the other relations of the b-value, were based on the monofractal property of faulting.

For the purpose of clarity of the discussion, we review the fractal relations of the b-value.

FRACTAL DEFINITION OF THE b-VALUE

In most cases the number of earthquakes occurring in a region within a specified length of time N with magnitude greater than m, satisfies the empirical magnitude-frequency relation (Gutenberg and Richter, 1954):

$$\log N = a - bm \tag{1}$$

where a and b are constants. As mentioned earlier the constant b, is generally referred to as the b-value; it describes the scaling property of earthquake size distribution. The scaling parameter b has been defined in terms of the fractal dimension by many authors as below:

(1) The moment of an earthquake is defined by

$$M = \mu \delta A \tag{2}$$

where μ is the shear modulus, A the rupture area and δ the average displacement on the fault break. The moment M can be related to its magnitude as

$$\log M = cm + d \tag{3}$$

where c and d are constants.
According to Kanamori and Anderson (1975), $c = 1.5$ and eqn. (2) can also be expressed in linear dimension of the fault break as

$$M = \alpha A^{3/2} \tag{4}$$

where α is constant.
Combining eqns. (1), (3) and (4) yields

$$\log N = -b \log A + \log s \tag{5}$$

where $\log s = bd/1.5 + \log a - b/1.5 \log \alpha$
From eqn. (5), we get

$$N = sA^{-b} \tag{6}$$

Equation (6) reduces to eqn. (7) for $A \sim r^2$

$$N = sr^{-2b} \tag{7}$$

The number-size distribution for a large number of objects is given by the fractal relation as

$$N = C/rD \tag{8}$$

where C is constant and N is the number of objects with a characteristic linear dimension greater than r.

Comparing eqns. (7) and (8) we get

$$D = 2b \tag{9}$$

So, the fractal dimension of seismic activity is twice the b-value. Figure 1 shows a plot between worldwide number of earthquakes per year with magnitude greater than 5 versus magnitude. However, for crystalline rocks the constant c (eqn. 3) is 3.0, hence D is equal to b.

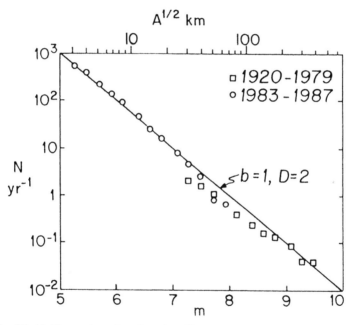

Fig. 1 Worldwide number of earthquakes N per year with magnitudes greater than m showing a power law with $b = 1$ and $D = 2$ (after Turcotte, 1986)

(2) Another way of using the moment-magnitude relation of earthquakes for fractal definition of b-value can be as follows:

Aki (1981) and King (1983) showed that eqn. (1) can be stated as:

$$N = Br\chi^b \tag{10}$$

where r is the rupture length, B is a constant related to a in eqn. (1) and $\chi = d/c$. The d is the dimension of the fault and c is a constant equal to 1.5 as before (Kanamori and Anderson, 1975).

Eqn (10) is equivalent to a fractal distribution, in which the fractal dimension D is given as

$$D = \chi^b \tag{11}$$

D is sometimes taken as the fractal dimension of fault length distribution in an area (e.g. Walsh and Waterson, 1992). This, however, implies that the rate of deformation along a fault is independent of the fault length. It has been shown that the power law scaling defined by the Gutenberg-Richter relation does not necessarily imply a fractal geometry as such can emerge solely from the dynamic or kinematic models of the earthquake process (Burridge and Knopoff, 1967; Bak and Tang, 1989). Hence, eqns. (10) and (11) are only statements of the fact that the size distribution of earthquakes is fractal. It gives no insight into the geometrical or dynamical origin of the b-value.

(3) A rigorous explanation for the origin of the b-value can be given in terms of the self-organised critical behaviour of the crust (Bak and Tang, 1989; Sornette and Sornette, 1989). The concept of self-organised criticality suggests that the crust is in a state of marginal stability, so that when perturbed from that state, it will evolve naturally back to the state of marginal stability. The energy input (strain) through the relative motion of tectonic plate is continuous but the energy loss is in a discrete set of events that satisfy the Gutenberg-Richter relation (Turcotte, 1994).

Sornette *et al.* (1991) and Lomnitz-Adler (1992) showed that if the scaling exponent of the self-organised dynamics of a population of earthquakes is b_u and the length distribution of faults has a fractal dimension D_l, then the observed b-value can be stated in terms of the capacity dimension D_c of the fault system, given that D_c has a linear relation with D_l (Nagahama and Yoshi, 1994) as,

$$b = b_u + \gamma D_c + \beta \tag{12}$$

where γ and β are constants. The b_u is defined in terms of the scaling exponent of strain distribution within the crust. Equation (12) is a theoretical relation that has to agree with observations. It is in particular difficult to determine the value of the parameter b_u for different regions, though Lomnit-Adler (1992) argued that it is a universal constant.

(4) It has recently been observed that there is linear correlation between the spatial distribution and the size distribution of earthquakes. For aftershock sequence, Guo and Ogata (1997) obtained a positive linear correlation between the b-value and the correlation dimension D_2. A negative correlation was obtained for the major earthquake sequence, however (Oncel *et al.*, 1996). The b-value-D_2 relation can therefore be stated as,

$$b = \mu D_2 + u \tag{13}$$

where μ and u are constants.The μ takes on a negative value for the major earthquake sequence and positive for the aftershock sequence.

The correlation dimension, defined by Grassberger and Procaccia (1983) measures the spacing or clustering of a set of points, which in this case is the earthquake epicentre. We note that if it is assumed that earthquakes occur only along faults, and that all faults are active , then the correlation dimension of an earthquake epicentre, taken over a sufficiently long period of time may be assumed to characterise the geometrical distribution of faults. Hence, eqns. (12) and (13) may be taken to be literally the same. Apart from the monofractal correlation of the *b*-value with the correlation dimension D_2, Hirabayashi *et al*. (1992) investigated the multifractal properties of earthquakes in California, Japan and Greece. They observed that the generalised dimension curve $(D_q - q)$ is of two distinct types. A steep type was observed during seismically active periods when large earthquakes occur, while a gentle type was seen during relatively quiet periods. This again suggests a correlation between the dynamical properties and the geometrical properties of seismicity. See Teotia (1999) in this volume for applications of mutifractal approach to earthquake data. Mabawonku and Dimri (1999) have applied the multifractal approach to microseismicity of North-East India.

APPLICATION TO KOYNA EARTHQUAKES

Seismicity in the Koyna India became prominent shortly after filling the Koyna dam in 1962 (Gupta and Rastogi, 1972). The dam is situated in the Peninsular shield of India which is believed to be relatively aseismic. The region has so far evidenced the largest reservoir related earthquake in the world with a magnitude of 6.2 on December 10, 1967. Seismic activity in the Koyna area has continued for the last 35 years. This intriguing continuous nature of seismicity is of considerable interest to seismologists around the world, as it provides an opportunity to understand plausible causative mechanisms for generation of such earthquakes.

In this paper we are not suggesting any mechanism for such a large number of earthquakes (with M greater than 4 earthquakes occurring every year and M equal and greater than 5 in 1968, 1973, 1980 and 1993-94), but we are attempting to estimate scaling parameters such as *b*-value, spatial and temporal correlation dimension from the distribution of the earthquakes, in order to ascertain a pattern in the seismicity of the region.

SPATIAL AND CORRELATION DIMENSION

We can estimate the spatial and temporal correlation dimension using the correlation integral method of Grassberger and Procaccia (1983) as

$$C(r) = 1/N^2 \sum_{i,j}^{N} \Theta\big((r - |x_i - x_j|)\big) \tag{14}$$

where Θ is the Heaviside function and counts how many pairs of points (x, x) fall below the interevent distance r. The correlation dimension D_2 is then defined by the scaling relation as

$$C(r) = r^{D_2} \tag{15}$$

The spatial and temporal correlation dimensions were determined from the slope of the curve between log $C(r)$ and log (r) using the least squares method as shown in Figs. 2 and 3 respectively for the Koyna area. The spatial and temporal correlation dimensions are 1.52 and 0.72 respectively. A value of 1.52 shows that although the epicentral distribution is localised, it is not particularly well clustered. The temporal correlation dimension, 0.72 shows that the time distribution of earthquakes is slightly clustered.

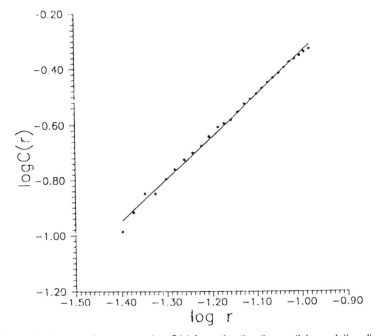

Fig. 2 A plot between log *r* versus log *C(r)* for estimating the spatial correlation dimension

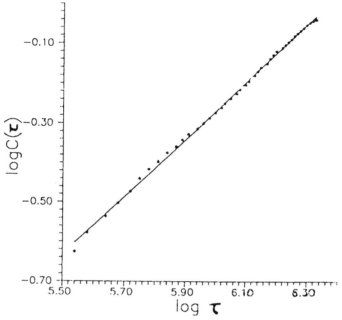

Fig. 3 The temporal correlation dimension using correlation integral method

The b-value was computed, and was 1.07. This value accords well with Gupta *et al.* (1972 a, b) who suggest that b-values for reservoir-associated seismicity are much higher than regional b-values. Mabawonku *et al.* (1998) have analysed in detail a relationship among b-values, spatial and temporal correlation dimensions of epicentral distribution of earthquake data and their variation in order to quantify the seismicity of the Koyna region. Also, see Srivastava (1999) in this volume.

CONCLUSION

Many definitions of b-value of the famous Gutenberg-Richter relation have been given in terms of fractal dimension. The b-value is half of the fractal dimension as found theoretically as well as from regional and worldwide earthquake records. The method to compute spatial and temporal correlation dimensions has been discussed. The b-value and spatial and temporal correlation dimensions have been computed for a number of earthquakes recorded in the Koyna area.

Acknowledgements: I am grateful to Dr. H.K. Gupta, Director NGRI, for his kind permission to publish this work. Thanks are due to Dr. Mabawonku, Dr. M. Ravi Prakash, Dr. P. Mandel and Mr. Abhey Ram for useful discussion and help during progress of the work.

REFERENCES

Aki K. 1981. A Probability Synthesis of Precursory Phenomena. In: *Earthquake Prediction, An International Review.* D.W. Simpson and P.G. Richards (eds.). Maurice Ewing. Ser., V. 4, ed. American Geophysical Union, Wash., DC.

Bak P and Tang C. 1989. Earthquake as a self-organized critical phenomenon. *J. Geophys. Res.,* **94**: 15635-15637.

Burridge R and Knopoff L. 1967. Model and theoretical seismicity. *BSSA,* **57**: 341-371.

Grassberger P and Procaccia I. 1983. Characterization of strange attractor. *Phys. Rev. Lett.,* **50**: 346-349.

Guo Z and Ogata Y. 1997. Statistical relations between the parameters of aftershock in time, space and magnitude. *J. Geophys. Res.* 102: 2857-2873.

Gupta HK and Rastogi BK. 1972. *Dams and Earthquakes.* Elsevier Sci. Publ., Amsterdam.

Gupta HK, Rastogi BK and Hari Narain. 1972a. Some discriminatory characteristics of earthquakes near Kirba, Kremasta, Koyna artificial lakes. *BSSA,* **63**: 493-505.

Gupta HK, Rastogi BK and Hari Narain. 1972b. Common features of reservoir associated activities. *BSSA,* **62**: 481-492.

Gutenberg B and Richter CF. 1954. *Seismicity of the Earth.* Princeton Univ. Press, NJ.

Hirabayashi T, Ito K and Yoshi. 1992. Multifractal analysis of earthquakes. *Pure & Applied Geophysics.* **138**: 591-610.

Kagan YY 1981. Spatial distribution of earthquakes: The three-point correlation function. *Geoph. JRAS,* **67**: 697-717.

Kagan Y Y and Knopoff L. 1980. Spatial distribution of earthquakes: The two-point correlation function. *Geoph. JRAS,* **62**: 303-320.

Kanamori H and Anderson D L. 1975. Theoretical basis of some empirical relations in seismology. *BSSA,* **65**: 1073-1095.

King G. 1983. The accommodation of large strains in the upper lithosphere of the earth and other solids by self-similar fault system: The geometrical origin of the *b*-value. *Pure & Applied Geophysics,* **121**: 761-815.

Lomnitz-Adler J. 1992. Interplay of fault dynamics and fractal dimension in determining Gutenberg-Richter's *b*-value. *Geoph. J. Int.,* **108**: 941-944.

Mabawonku AO and Dimri VP. 1999. Multifractal and multiscaling properties of the earthquakes with application to the microseismicity of North-East India. Submitted to *PAGEOPH.*

Mabawonku AO, Mandel P and Dimri VP. 1998. Self-organized fractal seismicity of reservoir induced earthquakes in Koyna, western India. Submitted to *PEPI.*

Nagahama H and Yoshi K. 1994. Scaling laws of fragmentation. In: *Fractals and Dynamic Systems in Geosciences,* pp. 25-36. J.H. Kruhl (ed.). Springer Verlag, Berlin.

Oncel AO Main I, Alptekin and Cowie, P. 1996. Spatial variation of the fractal properties of seismicity in the Anatonian fault zones. *Tectonophysics,* **257**: 189-202.

Sornette A and Sornette D. 1989. Self-organized criticality and earthquakes. *Europhy. Lett.* **9**: 197-202.

Sornette A, Davy P and Sornette D. 1991. Dispersion of *b*-value in Gutenberg-Richter law as a consequence of a proposed fractal nature of continental faulting, *GRL.* **18**: 897-900.

Srivastava HN. 1999. Chaotic dynamics and earthquakes. In: *Application of Fractals in Earth Sciences* (this volume) Dimri, V.P. (ed.).

Teotia SS. 1999. Multifractal analysis of earthquakes: An overview. In: *Application of Fractals in Earth Sciences*, (this volume) Dimri, V.P. (ed.).

Turcotte DL. 1986. A fractal model of crustal deformation. *Tectonophysics*, **132**: 261-269.

Turcotte DL. 1994. Crustal deformation and fractals: a review. In: *Fractals and Dynamic Systems in the Geoscience*. J.H. Kruhl (ed.), Springer-Verlag,. Berlin.

Walsh JJ and Waterson J. 1992. Population of faults and fault displacements and their effects on estimates *f* fault related regional extension. *J. Struc. Geology*, **14**: 701-712.

CHAOTIC DYNAMICS AND EARTHQUAKES

H.N. Srivastava*

INTRODUCTION: DETERMINISTIC CHAOS

A dynamical system whose equations and initial conditions are fully specified is called 'deterministic'. Solutions to deterministic equations become chaotic if adjacent solutions diverge exponentially in phase space. The concept of deterministic chaos has led in recent years to a new method of studying aperiodicity in complex systems such as earthquake occurrence. The unpredictability in such systems is attributed to non-linear interactions between the system and some unknown factors modelled as stochastic random noise.

Evolution of dynamical systems can be represented by trajectories in the state space from some initial condition. For a periodic system that develops deterministically, all trajectories initiated from different initial conditions stay on a low dimensional smooth topological manifold, called the attractor. These attractors are characterised by an integer dimension equal to the topological dimension of the submanifold. An important property of these attractors is that trajectories converging on them do not diverge. This guarantees long-term predictability of the system. It has been found for many dynamical systems that the trajectories stay on an attracting submanifold which is not topological. These submanifolds are

* Emeritus Scientist, B-1/52, Paschim Vihar, New Delhi 110 063, India.

called 'fractal' sets and are characterised by a dimension which is not an integer. The corresponding attractors are called 'strange' attractors. An important property of these attractors is the divergence of initially nearby trajectories. Thus, long-term predictability for these systems is not guaranteed.

In contrast, short-term constraints could be developed for attractors of low dimension in spite of unpredictability of a chaotic sytem in the long run. In other words, we can detect 'order' in chaos at least close to initial conditions. Determination of the dimension of an attractor sets a number of constraints that should be satisfied by a model used to predict the evolution of a system. The higher the value of the fractal dimension, the more complex the system. The fractal dimension also gives the number of independent parameters required for modelling the system. Their fractal nature brings out more details as they are increasingly magnified.

METHODOLOGY

Instead of having recourse to mathematical formulation of a non-linear system through differential equations, an alternative method is adopted in practice by replacing the state space by the so-called phase space, which is the co-ordinate space defined by the state variables of a dynamical system.

If the mathematical formulation of a system is not available, the state space can be replaced by the so-called phase space. The phase space may be produced using a single record of observable variable $x(t)$ from that system. The physics behind such an approach is that a single record from a dynamical system is the outcome of all interacting variables and thus information about the dynamics of that system should, in principle, be included in any observable variable.

It is assumed that variables present in the evolution of the system in question satisfy a set of n first-order differential equations:

$$\dot{x}_1 = f_1(x_1, x_2, \ldots x_n)$$

$$\dot{x}_2 = f_2(x_1, x_2, \ldots x_n)$$

$$\cdot$$
$$\cdot$$
$$\cdot$$

$$\dot{x}_n = f_n(x_1, x_2, \ldots x_n) \tag{1}$$

where, one dot indicates the first derivatives with respect to time. In such a case the co-ordinates of the state space are $(x_1, x_2, \ldots x_n)$. The above system of differential equations can be reduced to one highly non-linear differential equation of nth order, i.e.

$$x^{(n)} = f(x_1, x_2,\dots x_1^{(n-1)}) \tag{2}$$

Ruelle (1990) suggested that instead of a continuous variable $x(t)$ and its derivatives, it should be easier to work with $x_1(t)$, and the set of variables obtained from it by shifting its values by a fixed delay parameter λ. Let us now consider the phase space defined by the variables.

$$x_1(t),\ x_1(t + \lambda)\ \dots\ x_1[t+(n - 1)\lambda]$$

For a typical choice of t, these variables are expected to be linearly independent.

A set of N points on an attractor embedded in a phase space of n dimension is obtained from the time series.

$$x(t_1),\dots, x(t_N)$$

$$x(t_1 + \lambda),\dots, x(t_N + \lambda)$$

$$\cdot\qquad\qquad\cdot$$
$$\cdot\qquad\qquad\cdot$$
$$\cdot\qquad\qquad\cdot$$

$$x(t_1 + (n - 1)\lambda),\dots, x(t_N + (n - 1)\lambda)$$

Let

$$X_i[x(t_i),\ x(t_i + 1)\dots,\ x(t_i + (n - 1)\lambda)]$$

stand for a point of phase space.

The difference $|x_i - x_j|$ from the $N - 1$ remaining points is computed by the conventional Euclidean method for computing distance.

The correlation function of the attractor $C_m(r)$ is given by

$$C_m(r) = 1/N^2 \sum_{\substack{i,j=1 \\ i \neq j}} \theta(r - |x_i - x_j|) \tag{3}$$

for embedding dimension m, where θ is the Heaviside function

$$\theta(x) = 0, \qquad \text{if } x < 0$$

$$\theta(x) = 1, \qquad \text{if } x > 0$$

and r is distance.

The dimensionality d of the attractor is related to $C_m(r)$ by the relation

$$C_m(r) = r^D \ (\text{where } r \text{ is small}) \tag{4}$$

$$\log C_m(r) = D \log (r) \tag{5}$$

Hence, the dimensionality D of the attractor is given by the slope of the log $C_m(r)$ versus log (r).

Log $C_m(r)$ is plotted against log (r). The slope of the scaling region is obtained for various embedding dimensions. As we increase the embedding dimension m, the slope saturates to a limiting value, which is considered a fractal dimension of the strange attractor. The delay time is slowly increased until the same fractal dimension is obtained for two consecutive delay times.

Lyapunov exponent

We consider a system described by N ordinary differential equations.

$$\mathrm{d}x_i / \mathrm{d}t = F_i (x_i, \dots x_N), i = 1 \dots N \tag{6}$$

The solution space for this problem conceptually follows solutions that start within a hypersphere of radius r. As the solution evolves, the hypersphere is deformed into a hyperellipsoid with principal axes $\varepsilon_i(t)$. The Lyapunov exponent is

$$= \lim(t \to \infty) \lim (r \to 0) \{1/t \; \varepsilon_i(t)/r\} \tag{7}$$

If all $\lambda_i \le 0$, all solutions that start with initial conditions close to each other will converge, i.e. there is no sensitivity to initial conditions. But if just one λ_i is positive, the nearby solutions will diverge, i.e., there will be extreme sensitivity to the initial conditions.

The growth in uncertainty in time t is given by

$$N = N_0 e^{\lambda t} \tag{8}$$

where N_0 is initial condition and λ is related to the concept of entropy in information theory and also related to another concept, i.e., the Lyapunov exponent, which measures the rate at which nearby trajectories of a system in phase space diverge. The Lyapunov exponent λ of the time series is found using the method developed by Wolf *et al.* (1985). Positive Lyapunov exponent is suggestive of the chaotic nature of earthquakes, while the non-chaotic system is characterised by a negative Lyapunov exponent.

RESULTS AND DISCUSSIONS

Koyna reservoir

We have considered three periods for Koyna region as follows:
  (i) Period I covers data from February 1967 to December 1981. It includes the largest foreshock of magnitude 5.0 during September

1967, main shock of December 1967 and its immediate aftershocks of the same month, besides earthquakes of magnitude 5.0 during October 1973 and September 1980.

(ii) Period II includes data from January 1968 to September 1973 and is a part of period I. This period begins a few days after the occurrence of the main earthquake and ends a few days prior to the earthquake of 1973.

(iii) Period III covers the data from January 1968 to August 1980. It is a part of period I but also includes period II. This period begins a few days after the occurrence of the main earthquake and ends a few days prior to the earthquake of September 20, 1980. The period contains the earthquake of October 17, 1973.

The result for period I is shown in Figs. 1 and 2. It may be noted that the slopes for days 4 and 6 converge to 4.2 as the embedding dimension is increased. We get the fractal dimension of the attractor as 4.4 for period II. The results for period III gives fractal dimension of the attractor as 4.4. It may be mentioned that care has been taken to avoid spurious results being obtained by keeping the number of earthquakes N for each period such that the criterion $2 \log N \geq D$ is satisfied. Here D is the fractal dimension. In view of consistency in the fractal dimension in the Koyna region, Srivastava *et al.* (1994) suggested it as a new measure of seismotectonics. Srivastava *et al.* (1995) extended similar studies to Aswan and Nurek reservoirs and found a strange attractor dimension of 3.8 and 7.2 respectively. Since Koyna and Aswan reservoirs are located in shield regions, compared to Nurek reservoirs in a complex zone, the difference in predictability of earthquakes is highlighted.

North-East India

The strange attractor dimension was 6.1 within the epicentral distance of 220 and 440 km radii around Shillong. The former area covers the Shillong massif bounded by faults, namely Dawki, Haflong-Disang and Um Ngot lineament. The larger area included the zone of the Indo-Burman ranges (with intermediate depth earthquakes), Dhubri faults and other lineaments. The positive values of the largest Lyapunov exponent were found to be 0.064 and 0.057 respectively for the two areas, suggesting sensitivity to initial conditions of the supposed earthquake dynamics. The values of the strange attractor dimension around Shillong are broadly in agreement with that in the Hindukush region (Bhattacharya and Srivastava, 1992) but less compared to Himachal Pradesh and Nurek regions (Tadjik) (Bhattacharya, 1990; Bhattacharya *et al.*, 1995). The minimum and the maximum number of parameters for earthquake predictability are 7 and 12 respectively around Shillong plateau and environs.

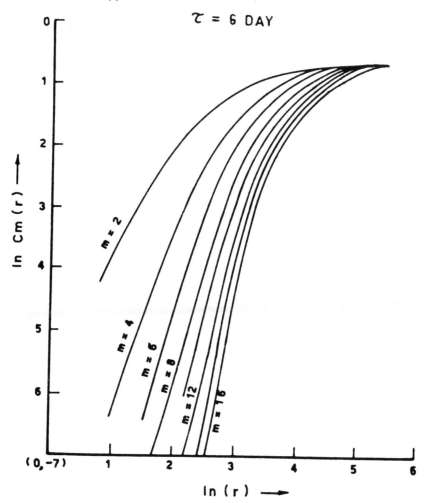

Fig. 1 Koyna earthquake (Feb. 1967 - Dec. 1981) distance dependence of the correlation function

The above results suggest self-organised criticality in the region, according to which the system is in the critical state. The fractal frequency magnitude statistics may be considered to be applicable (Turcotte, 1992). We may therefore compare the strange attractor and the fractal dimension for the Indian region.

The value of b in the Gutenberg-Richter frequency magnitude relationship for north-east India was found as nearly 1 for two different periods of observation, namely 1970-1972 and 1963-1972 (Chaudhury and Srivastava, 1976). Srivastava and Dattatrayam (1988) reported a similar b value of 0.95 in the region based on the data during the years 1964-1987.

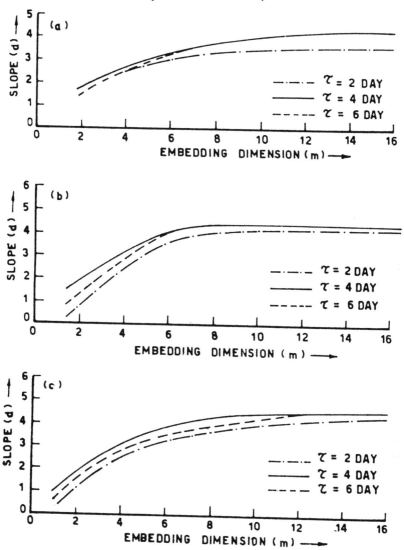

Fig. 2 Dimensionality d of the attractor as a function of embedding dimensions

For Koyna region also, we get a *b* value of 0.97 for the period 1967-1987. For the Himalayan region a *b* value of the same order has been reported. By taking the fractal dimension as twice the *b* value, we hardly notice any difference between interplate and intraplate earthquakes. On the other hand, a smaller value of strange attractor dimension of 4.5 was found for the Koyna region (Srivastava *et al.*, 1994) compared to north-east India or Himalaya-Hindukush region (Bhattacharya, 1990; Bhattacharya and

Srivastava, 1992). Thus, we note a marked distinction in the chaotic dynamics between intraplate Hindukush-Himalaya (including north-east India) vis-à-vis interplate Koyna region.

PRECURSORY SEISMICITY PATTERNS AND STRANGE ATTRACTOR DIMENSION

Precursory seismicity patterns are delineated by including earthquakes originating on the same fault as well as neighbouring regions, which may be associated with different tectonic features (Srivastava *et al.*, 1987; Gupta and Singh, 1989; Srivastava and Rao, 1991). A question arises as to why 'distant' earthquakes located on different fault systems should be included to study seismicity patterns. We present applications of chaotic dynamics through the case study of seismicity pattern preceding the Indo-Burma border earthquake of August 1988 (M = 7.0), which had a focal depth of about 100 km. As mentioned earlier, several studies have shown that the earthquakes in this region possess mixtures of thrust, strike slip and normal faulting (Chen and Molnar, 1990; Chandra, 1978; Chaudhury and Srivastava, 1976; Ichikawa *et al.*, 1972).

Srivastava and Rao (1991) took all the earthquakes (shallow or inter-mediate depth) within a radial distance of 300 km from the epicentre of the impending earthquake of August 1988 and found a well-defined clus-tering of seismic activity from 1981 to 1987. Another methodology was adopted by Gupta and Singh (1989), based on identifying swarms and quiescence before the main earthquake over a slightly larger grid size of about 4 to 6° (about 3 to 4° radius). Both the seismicity patterns were based on the U.S.G.S. catalogue of earthquakes which had the same detection capability during the period 1963-1988. They included earth-quakes of all the focal depths extending down to 100 km or more. Since the focus of the earthquake of August 1988 was in the upper mantle, further questions arise about including crustal earthquakes to delineate the seismicity pattern due to different types of deformation at the surface compared to that at depth, particularly near the Indo-Burma border. As discussed earlier, the strange attractor dimension remained the same for small (220 km) as well large (440 km) areas, implying similar chaotic dynamics, thus providing justification to include earthquakes in near as well as far field regions to delineate the seismicity pattern in this zone of complex tectonics.

Table 1 Strange attractor Dimension vis-à-vis stress drop in great earthquakes (M ≥ 8)

Region	Strange attractor dimension	Stress drop (bars)	b value	Reference
(i) Koyna Region	4.4	0.97		
(ii) Himalaya				
(a) North-East India	6.1	140, 213, 398 (15 Aug., 1950, M = 8.5)	0.95	Tandon and Srivastava (1974) (using different methods)
(b) Central India Nepal border	Not determined	275 (14 Jan. 1934, M = 8.3)	0.65	Chen and Molnar (1977)
(c) Himachal Pradesh	9.8	Not known	0.85	Bhattacharya (1990)
(d) Hindukush	6.9	Not known	1.2	Bhattacharya and Srivastava (1992)
(iii) Cocos-South American plate boundary (near Mexico)	Not determined	25 (19 Sep. 1985, M = 8.1)		UNAM Seismology Group (1986)
(iv) Pacific-Eurasian plate boundary (near Japan)	3.2	33 (2 Mar 1933, M= 8.7)		Kanamori (1971) Pavlos et al. (1994)

STRESS DROP VERSUS STRANGE ATTRACTOR DIMENSION

Regional variations in stress drop have importance in seismic risk analysis due to their dependence on the level of acceleration produced by an earthquake. Large differences in the stress drop have been reported for great earthquakes (M = 8.0) in continent collision Himalayan zone vis-à-vis subduction zones (Srivastava, 1988). Near the Indo-Eurasian plate boundary, the stress drops are higher compared to subduction zones of the Pacific and Cocos plate boundaries. It may be noted from Table 1 that the strange attractor dimensions in these regions follow similar trends of variations, suggesting the influence of fault dynamics on the stress drop. However, such differences in stress drop become less marked for earthquakes of magnitude 6. Also, the stress drop estimates become worse with increasing number of fault segments using kinematic analysis based on asperity and barrier models (Rudnicki and Kanamori, 1981). The fault dynamics inferred through the deterministic approach may therefore provide better insight into seismic risk through the consistency in the strange attractor dimension, compared to the stress drop measurements whose values vary with the magnitude of the earthquake, nature of faulting and inherent errors in the computational procedure.

REFERENCES

Beltrami H and Marschal JC. 1993. Strange seismic attractor? *PAGEOPH*, **141**: 72-82.

Bhattacharya SN. 1990. Quantifying chaos in earthquake occurrence in Dalhousie-Dharamsala area in north-west Himalaya. Nat. Symp. on Recent Advances in Seismology and Their Applications, Bangalore (Abstract).

Bhattacharya SN and Srivastava HN. 1992. Earthquake predictability in Hindukush region using chaos and seismicity pattern. *Bull. Ind. Soc. Earthq. Tech.* **29**: 23-25.

Bhattacharya SN, Sinha Ray KC and Srivastava HN. 1995. Large Fractal Dimension around Nurek Reservoir. *Mausam*, **46**: 187-192.

Chandra U. 1978. Seismicity, Earthquake Mechanisms and Tectonics Along the Himalayan Mountain Range and Vicinity. *Phys. Earth Planet. Int.* **16**: 109-131.

Chaudhury HM and Srivastava HN. 1976 Seismicity and Focal Mechanism of Earthquakes in North-East India, *Annal. Geofis.*, **28**: 382-391.

Chen WP and Molnor P. 1990. Source Parameters of Earthquakes and Intraplate Deformation Beneath the Shillong Plateau and Northen Indo-Burman Ranges. *J. Geophys. Res.* **95**: 12527-12552.

Grassberger P and Procaccia I. 1993. Characterization of strange attractors. *Phys. Rev. Lett.* **50**: 346-349.

Gupta HK and Singh HN. 1989 Earthquakes Swarms Precursory to Moderate to Great Earthquakes in the North-East Indian Region. *Tectonophysics*, **167**: 285-298.

Horowitz FG. 1989. A strange attractor underlying Parkfield Seismicity ? *EOS*, **70**: 1390.

Ichikawa M, Srivastava HN and Drakopoulos JC. 1972. Focala Mechanism of Earthquakes Occurring in and Around the Himalayan and Burmese Mountain Belt. *Meteor. and Geophys. (Tokyo)*, **23:** 149-162.

Julian BR. 1990. Are earthquakes chaotic ? *Nature*, **345:** 481-482.

Pavlos GP, Karakatsanis L, Latoussakis LB, Dialetis D and Papaioannov G. 1994. Chaotic analysis of a time series composed of seismic events recorded in Japan. *Int. J. Bifur. and Chaos*, **4:** 87-98.

Rudnicki JW and Kanamori H. 1981. Effects of Fault Interaction on Moment, Stress Drop and Strain Energy Release. *J. Geophys Res.*, **86:** 1785-1793.

Ruelle D 1990. Deterministic Chaos: The Science and the Fiction (The Claude Bernard Lecture). *Proc. R. Soc. London (Maths. Phys. Sci.)* **A424:** 241-248.

Ruelle D 1990. Deterministic Chaos: The Science and the Fiction (The Claude Bernard Lecture). *Proc. R. Soc. London (Maths. Phys. Sci.)* **A427:** 241-248.

Sornette A, Dubois J, Cheminee JL and Sornette D. 1991. Are sequences of volcanic eruptions deterministically chaotic? *J. Geophys. Res.* **96:** 11931-11945.

Srivastava HN and Rao PCS. 1991. Seismicity Pattern Associated with Earthquakes of August, 1988 near Manipur-Burma and Bihar-Nepal Regions. *Bull. Ind. Soc. Earth. Tech.* **28:** 13-22.

Srivastava HN. 1988. The Mexican Earthquake of 1985 vis-à-vis Great India Earthquakes: Field and Seismological Aspects. *Mausam*, **39:** 149-158

Srivastava HN, Dube RK and Hans Raj. 1987. Space and Time Variation in Seismicity Patterns Preceding Two Earthquakes in the Himachal Pradesh, India. *Tectonophysics*, **113:** 69-77.

Srivastava HN, Bhattacharya SN and Sinha Roy KC. 1994. Strange attractor dimension as a new measure of seismotectonics around Koyna reservoir, India. *Earth Planet Sci. Lett.*, **124:** 57-61.

Srivastava HN, Bhattacharya SN and Sinha Ray KC. 1996. Strange attractor characteristics of earthquakes in Shillong and neighbourhood. *Geophys. Res. Lett.*, **23:** 3519-3522.

Srivastava HN, Bhattacharya SN, Sinha Ray KC, Mahmoud SM and Yunga S. 1995. Reservoir associated characteristics using deterministic chaos in Aswan, Nurek and Koyna reservoirs. *PAGEOPH*, **145:** 205-207.

Turcotte DL. 1992. *Fractals and Chaos in Geology and Geophysics*. Cambridge Univ. Press, New York.

Wolf A, Swift JB, Swinney LH and Vastano JA. 1985. Determining Lyapunov exponents from a time series. *Physica*, **16D:** 285-317.

MULTIFRACTAL ANALYSIS OF
EARTHQUAKES: AN OVERVIEW

S.S. Teotia*

INTRODUCTION

Fracture exhibits a fractal structure over a wide range of fracture scales, i.e., from the scales of microfractures to megafaults (Brown and Scholz, 1985; Scholz and Aviles, 1986; Okubo and Aki, 1987; Aviles *et al.*, 1987; Hirata *et al.*, 1987). In such systems the number of fractures that are larger than a specified size are related by power law to the size. The physical laws governing the fractal structures are scale invariant in nature, since the occurrence of earthquakes is causally related to the fractures which have a fractal structure in their space, time and magnitude distributions. The fractal structures may have either homogeneous or multiscaling fractal sets. Fractal sets having multiscaling are heterogeneous fractal sets and are called multifractal sets. Most fractals in nature are said to be heterogeneous (Stanley and Meakin, 1988; Mandelbrot, 1989). Such fractals are characterised by generalised dimension D_q, where q takes the value of 0, 1, 2. Recent studies have shown that many natural phenomena, such as spatial distributions of earthquakes, fluid turbulence, are heterogeneous fractals (Frisch and Parisi, 1985; Hirata and Imoto, 1991).

* Department of Geophysics, Kurukshetra University, Kurukshetra 136 119, India.

The heterogeneity and multiscaling of the fractal structure of spatial distribution of an earthquake in a region is related to the heterogeneous distribution of seismicity. Seismologists have been using multifractal concepts to investigate heterogeneity in seismicity in various seismic regions of the world. The spatial distribution of earthquakes is found to change before and after large earthquakes. This change is reflected in the generalised dimension D_q or D_q spectra. Therefore, the temporal variation of D_q and D_q spectra may be used to study the changes in seismicity structure before the occurrence of large earthquakes and hence multifractal study holds promise in forecasting earthquakes.

THEORY

Earthquakes are considered point processes. A fractal set presents a self-similar organisation, without characteristic length. Fractal objects are characterised by their fractal dimension (Mandelbrot, 1975). Besides simple fractal sets, more complex ones have been observed, e.g. clusters within other clusters. In such cases the structure cannot be described by a single fractal dimension constituting a spectrum of dimensions. In general, there is an infinite set of fractal dimensions and all of them are necessary to define the multifractals.

Capacity dimension (D_c)

The complete space containing the fractal set is covered with boxes of size r. The capacity dimension takes into account only the fact that the boxes are occupied or not. The capacity dimension is:

$$D_c = \lim_{r \to 0} \log N(r)/\log (1/r) \tag{1}$$

where $N(r)$ is the number of non-empty boxes of size r needed to cover the fractal set. The capacity dimension is also known as the Hausdorff dimension. It is often called the Box Counting Method.

Information dimension (D_i)

This takes into account a new information, i.e. the number of points which are in each non-empty i^{th} box of size r. The information dimension D_i is defined for probabilistic distribution as:

$$D_i = \lim_{r \to 0} \sum_{i=0}^{N} P_i(r) \log P_i(r)/\log r \tag{2}$$

where $P_i(r)$ is the probability of occupation of the ith box of size r, N is the number of boxes of size r. The information dimension does not take into account the way in which points are distributed within each box. D_c and D_i take the same value if all the boxes have equal probability. However, for the general case, the information dimension is always a lower boundary to the Hausdorff dimension (Grassberger and Procaccia, 1983).

Correlation dimension (D_{COR})

Evolutionary geophysical processes can be described by trajectories in the state space. Each trajectory in the state space represents the evolution of the system from some initial conditions. These physical systems can exhibit an attractor, as trajectories develop from different initial conditions and the system will eventually converge and stay on submanifolds of the total available space. This submanifold 'attracts' the trajectories and hence is called an attractor. A useful way to quantify the attractor is to determine its correlation dimension D_{COR}. Attractors which display chaotic and turbulent behaviour were termed strange attractors (Rulle, 1990; Takens, 1981). This fractal dimension was first introduced by Grassberger and Procaccia (1983) for a smoothed dynamical system and later applied for a set of points in space. The correlation dimension is defined by:

$$D_{COR} = \lim_{r \to 0} \log C(r)/\log r \tag{3}$$

where $C(r)$ is the correlation function, given by

$$C(r) = 1/N^2 \sum_{\substack{i,j=1 \\ i=j}}^{N} H\left(r - \left|X_i - X_j\right|\right) \tag{4}$$

where, r is the scaling radius, N the total number of data points within a search region in a certain time interval (also called the sample volume) and $|X_i - X_j|$ the distance between two points X_i and X_j. H is the Heaviside function.

The relationship between the correlation dimension and the cumulative correlation function is based on the power law, i.e. $C(r) = r^D$. A determination of correlation dimension is found by plotting $C(r)$ versus r on a log-log graph. The region in which the power law is obeyed is called the scaling region. The slope (which is an estimate of correlation dimension) is formed by fitting a least squares line in the scaling region. The correlation dimension depends on the number of data points used in the analysis of dynamical systems.

Generalised dimension (D_q)

Analogous questions arise in the dynamical system theory when one is interested in describing probabilities of visiting a given part of a strange attractor. For this purpose, the multifractal approach was suggested (Halsey *et al.*, 1986), which involves a whole spectrum of non-integer dimensions of the set under consideration. Mandelbrot (1989) pointed out that most fractals in nature and in dynamical systems are not uniform and symmetric but are known to be heterogeneous. For such systems, a unique fractal dimension is not sufficient to characterise them. Such dynamical systems possess a multifractal nature and are characterised by the generalised dimension D_q or the $f(\alpha)$ spectrum (Halsey *et al.*, 1986). The generalised dimension D_q is more useful for the comprehensive study of heterogeneous fractals. Therefore, the generalised dimension D_q is a parameter representing the complicated fractal structure or multiscaling nature. The general methods for calculating D_q are the fixed-mass method, the fixed-radius method and the box counting method (Grassberger *et al.*, 1988; Baddi and Broggi, 1988; Greenside *et al.*, 1982). These methods work well provided the number of data points is very large. The extended Grassberger and Procaccia method (Grassberger and procaccia, 1983; Pawelzik and Schuster, 1987) is used to recover the dimension from a time series. Its formula is:

$$\log C_q(r) = D_q \log r \ (r \rightarrow 0) \tag{5}$$

where $C_q(r)$ is the generalised correlation function given by

$$C_q(r) = \left\{ 1/N \sum_{j=1}^{N} \left[(1/N) \sum_{i=j}^{N} H\left(r - |X_i - X_j|\right) \right]^{q-1} \right\}^{1/q-1} \tag{6}$$

Capacity, information and correlation dimensions are special cases of generalised dimension for $q = 0, 1, 2$. In heterogeneous fractals the D_q decreases with q and D_q is independent of q in case of homogeneous fractals (see Fig. 1).

MULTIFRACTAL STUDIES OF EARTHQUAKES

Studies by seismologists have shown that the occurrence of earthquakes is fragmented, contains gaps, and has heterogeneous distribution both in time and space (see Figs. 2 and 3). Therefore, time and space distributions of earthquakes are fractals, have fractal structure within a certain range of scale and can be measured by fractal dimensions. The basis of

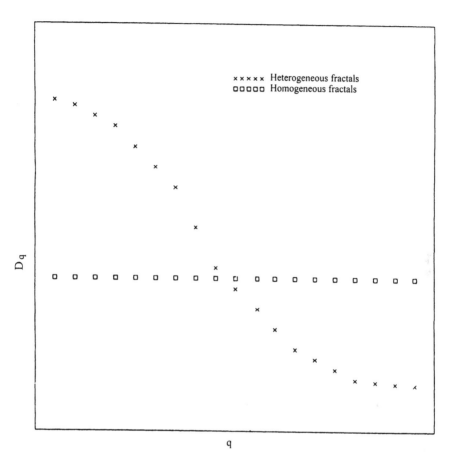

Fig. 1 D_q decreases with q for heterogeneous fractals whereas D_q is independent of q for homogeneous fractals

fractal structure is the self-similarity or the scale-invariant characteristics. Seismologists have suggested that multifractal analysis is more useful for the study of heterogeneous fractals of earthquakes. The spatiotemporal variation in the occurrence of earthquakes may be studied in the framework of multifractals, i.e, by studying the (i) $f(\alpha)$ spectrum or generalised dimension D_q of various seismogenic regions and (ii) temporal variation of D_q spectra.

REGIONAL SEISMICITY STUDIES

To understand the various seismogenic regions in terms of spatial

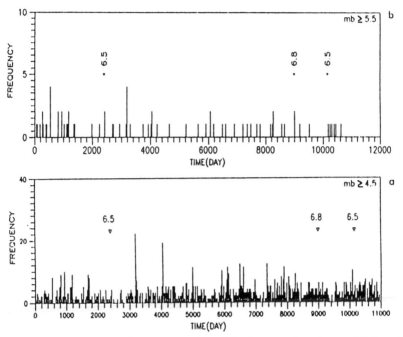

Fig. 2 Temporal change in frequency number of Himalayan earthquakes per 30 days. (a) Frequency for magnitude threshold *mb* ≥ 4.5. (b) Frequency for magnitude threshold *mb* ≥ 5.5 (after Teotia, Ph.D. thesis, Kurukshetra Univ.)

intermittency of seismicity as well as patches of concentrated and rarified seismicity, the functions $f(\alpha)$ spectrum and generalised dimension D_q are used. Multifractal properties of the epicentre, hypocentre distribution and energy distribution using the generalised dimension have been studied for California, Japan and Greece (Hirabayashi *et al.*, 1992).

Earthquake prediction studies

The D_q versus q curve is termed the D_q spectrum. The total time-series can be divided into various subsets having the same data volume. Variation in the D_q spectrum from one subset to another reveals the change in seismicity pattern, which may be attributed to precursory changes in a given region. In general, the value of D_q is larger for low values of q and decreases with increasing value of q. However, the rate of decrease (slope) may remain the same or change from one subset to another. The high rate of decrease corresponds to a steep type of D_q spectrum and a low rate corresponds to a gentle type of D_q spectrum (see Figs. 4 and 5). D_q spectra are expected to be sensitive to temporal changes in seismicity

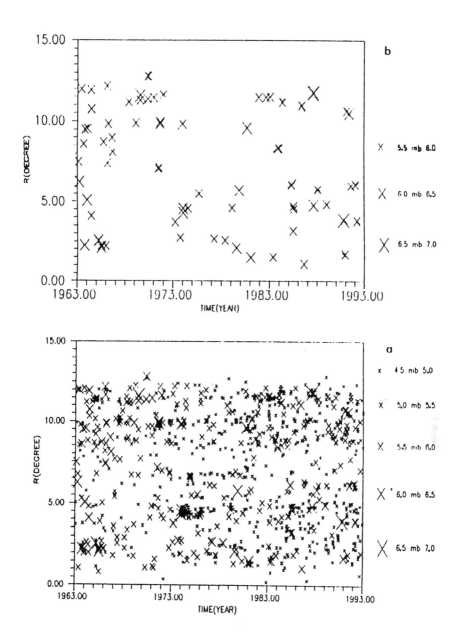

Fig. 3 Spatiotemporal pattern of seismicity in $R - t$ domain having different magnitude threshold: (a) Magnitude threshold $mb \geq 4.5$ (b) Magnitude threshold $mb \geq 5.5$. R is the distance from reference point to all 1099 earthquakes occurring in Himalayan region (after Teotia, Ph.D. thesis, 1995)

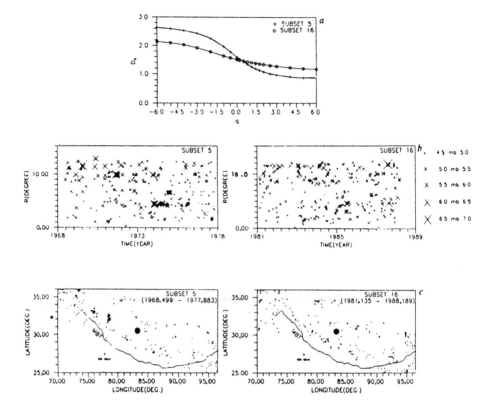

Fig. 4 (a) D_q spectra of subset 5 and subset 16; (b) spatiotemporal pattern of seismicity in $(R - T)$ domain; (c) epicentral distribution of earthquakes occurring in different time windows in subset 5 and subset 16 (Teotia *et al.*, 1997)

patterns (i.e., clusters and gaps) before large earthquakes. The slope of D_q spectra becomes gentle for extended distribution of earthquakes and steep for clustered distribution of earthquakes. Such studies have been done by various researchers in various seismic regions of the world (Hirata and Imoto, 1991; Hirabayashi *et al.*, 1992; Li *et al.*, 1994; Legrand *et al.*, 1996; Teotia *et al.*, 1997).

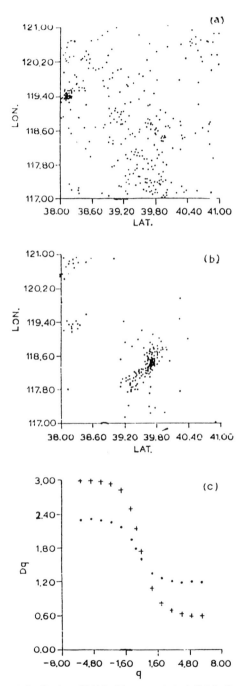

Fig. 5 (a) Extended distribution, *N*-400; (b) concentrated distribution, *N*-400; (c) D_q spectra of (a) and (b) (after, Li *et al.*, 1994)

REFERENCES

Aviles CA, Scholz CH and Boatwright J. 1987. Fractal analysis applied to characteristic segments of the San Andreas fault. *J. Geophys. Res.*, **92**: 331-344.

Baddi R and Broggi G. 1988. Measurement of the dimension spectrum $f(\alpha)$: Fixed-mass approach. *Phys. Lett.* **A131**: 339-343.

Brown SR and Scholz CH. 1985. Broad bandwidth study of the topography of the natural surfaces. *J. Geophys. Res.*, **90**: 12,575-12,582.

Frisch U and Parisi G. 1985. In: *Turbulence and Predictability in Geophysical Fluid Dynamics and Climate Dynamics*. p. 84 M. Ghil *et al.* (eds.) North-Hollan, Amsterdam.

Grassberger P. 1983. Generalised dimension of strange attractor. *Phys. Lett.*, **97A**: 227-230.

Grassberger P and Procaccia I. 1983. Measuring the strangeness of strange attractors. *Physica* **9D**: 189-208.

Grassberger P, Baddi R and Politi A. 1988. Scaling laws for invariant measures on hyperbolic attractors. *J. Stat. Phys.* **51**: 135-178.

Greenside HS, Wolf A, Swift J and Pignataro T. 1982. Impracticality of a box counting algorithm for calculating the dimensionality of strange attractors. *Phys. Rev.* **A25**: 3453-3459.

Halsey TC, Jense MH, Kadanoff LP, Procaccia I and Shariman BI. 1986. Fractal Measures and Their Singularities: The Characterization of Strange Sets. *Phys. Rev.* **A33**: 1141-1151.

Hirabayashi T, Ito K and Yoshii T. 1992. Multifractal analysis of earthquakes. *PAGEOPH*, **138**: 591-610.

Hirata T and Imoto M. 1991. Multi-fractal analysis of spatial distribution of microearthquakes in the Kanto region. *Geophys. J. Int.* **107**: 155-162.

Hirata T, Satoh T and Ito K. 1987. Fractal structure of spatial distribution of microfracturing in rocks. *Geophys J.R. Astro. Soc.* **90**: 360-374.

Legrand D, Cisternas, A and Dorbath A. 1996. Multifractal analysis of the 1992 Erzincan aftershock sequence. *Geophys. Res. Lett.*, **23**: 933-936.

Li D, Zheng Z and Wang B. 1994. Research into the multifractal of earthquake spatial distribution. *Tectonophysics*, **233**: 91-97.

Mandelbrot BB. 1975. Stochastic model for the earth's relief, the shape and fractal dimension of coastlines and the number-area rules for island. *Proc. Nat Acad. Sci. USA*, **72**: 3825-3828.

Mandelbrot BB. 1989. Multifractal measure, especially for the Geophysicist. *Pure and Appl. Geophys.*, **131**: 5-42.

Okubo PG and Aki K. 1987. Fractal geometry in the San Andreas fault system. *J. Geophys. Res.*, **92**: 345-355.

Pawelzik K and Schuster HG. 1987. Generalised dimension entropies from a measured time series. *Phys. Rev.* **35**: 481-484.

Ruelle D. 1990. Deterministic chaos: the science and the fiction. *Proc. R. Soc. Land. A.*, **427**: 241-248.

Scholz CH and Aviles CH. 1986. The fractal geometry of faults and faulting. In: *Earthquake Source Mechanics*. Das *et al.* (eds.). (AGU, Wash. DC) **37**: 147-155.

Stanley HE and Meakin P. 1988. Multifractal phenomena in physics and chemistry. *Nature*, **335**: 495-499.

Takens F. 1981. Detecting strange attractors in turbulence. *Lecture Notes in Mathematics*, (Springer, Heidelberg-New York), p. 898.

Teotia SS, Khattri KN and Roy PK. 1997. Multifractal analysis of seismicity of Himalayan region. *Current Science*, **73**: 359-366.

Teotia SS. 1995. *Multifractal Analysis of Earthquakes in Himalaya*. Ph.D thesis, Kurukshetra University, Kurukshetra.

APPLICATION OF FRACTALS IN THE STUDY OF ROCK FRACTURE AND ROCKBURST-ASSOCIATED SEISMICITY

K. Shivakumar* and M.V.M.S. Rao**

INTRODUCTION

Many natural systems and processes are scale-invariant or fractal in character (Mandelbrot, 1982) and geology and geophysics are no exception. Several studies in recent years have shown that the fracture process of rock at the laboratory scale or during large earthquakes has self-similarity (fractal structure) in space, time and magnitude distributions as expressed by the fractal dimension, D, Omori's exponent, P, and the b-value respectively (Kagan and Knopoff, 1980; Hirata et al., 1987; Hirata, 1989; Aki, 1965). This fundamental and pervasive scale-invariance of rock fracture phenomenon has paved the way for the application of damage mechanics theories and fractal techniques to investigate in depth the failure behaviour of rock at the laboratory scale (Hirata et al., 1987; Lei et al., 1992; Shivakumar et al., 1994; Rao, 1996) as well as in underground mines where it is known as rock burst (Coughlin and Kranz, 1991; Xie and Pariseau, 1993; Shivakumar et al., 1996; Shivakumar and Rao, 1997).

* National Institute of Rock Mechanics, Kolar Gold Fields, Karnataka, India.
** National Geophysical Research Institute, Hyderabad, India.

A rockburst is a localised macrofracture of brittle rock in an underground opening with an intermediate scale (m) between microfracturing (cm) and an earthquake (km). The physical process experienced by a rockburst is a damage evolution process. Rockbursts constitute an important set of mine-induced seismic events. Hazardous rockbursts are reported to occur in the deep mines of the Kolar Gold Fields, causing death and injury to the mine workers and damage to the underground mine structures (Xie and Pariseau, 1993; Shivakumar and Rao, 1997; Rao et al., 1994; Srinivasan et al., 1997). Microseismic and seismic investigations of 'suspect areas' and rockbursts in deep mines have advanced well in recent years following the application of fractal techniques for determination of the correlation dimension, D_c (Coughlin and Kranz, 1991; Xie and Pariseau, 1993, Shivakumar et al., 1996; Shivakumar and Rao, 1997). Further, laboratory studies of the acoustic emissions (AE) associated with the fracture and failure of rock under simulated field conditions have clearly established the fact that the strain energy release increases exponentially with decrease in the correlation dimension, D_c and the seismic b-value as the impending failure approaches (Hirata et al., 1987; Lei et al., 1992; Rao, 1996). Thus, the application of fractal techniques leads to a better description and interpretation of the seismic and microseismic character of rockbursts in deep mines, and also can contribute to the accuracy and reliability of microseismic and acoustic emission techniques for the prediction and control of rockbursts.

The application of fractal techniques to rockburst seismicity problems of the Kolar gold mines, India was undertaken recently at the National Institute of Rock Mechanics (Shivakumar et al., 1996; Shivakumar and Rao, 1997; Rao et al., 1994; Srinivasan et al., 1997). The stochastic self-similar concept of the spatial distribution of the hypocentres of seismic, microseismic and acoustic emissions of the burst-prone rocks of the Kolar gold mines was investigated in detail (Shivakumar et al., 1994; 1996; Shivakumar and Rao, 1997). This became possible by short-listing and analysing the qualifying data acquired by a seismic network (7 km × 3 km × 3 km), microseismic network (500 m × 200 m × 500 m), and AE network (100 mm long and 50 mm diameter) at the field sites and during the laboratory experiments respectively. The important results obtained are summarised and discussed below.

FRACTAL MEASUREMENT METHODS

Number-radius relationship

The AE or microseismic or seismic event locations construct a spatial distribution of a point set in which a point corresponds to a cracking surface or volume element in physical space. Thus the fractal dimension of the damage evolution process at any given scale can be measured directly from the distribution of the point set (Xie and Pariseau, 1993). Considering a sphere with radius r, the total number of events inside this sphere over the distribution can be counted and denoted by $M(r)$. A set of data $M(r_i)$ associated with different radii r_i (i = 1, 2, 3, ...) can be obtained from fractal geometry (Xie and Pariseau, 1993). There is a relation between $M(r_i)$ and r_i in the form $M(r) = r^1$ for the line distribution of point set, $M(r) = r^2$ for the plane distribution, and $M(r) = r^3$ for the 3-D (or volume) distribution, and

$$M(r) = r^D \qquad (1)$$

for a fractal distribution with fractal dimension D. Equation (1) is also called the number-radius relation and the fractal dimension, D, is called the clustering dimension, which is equal to the slope of the log $M(r)$ – log(r) plot. In this fractal measurement, the centre point of the spheres with different radii r_i is chosen as the mass centre of the distribution.

Correlation exponent method

Spatial self-similarity can be demonstrated by examining the distribution of distances between pairs of points in a data set over a range of distances. This has been done on the earthquake scale (Kagan and Knopoff, 1980; Hirata, 1989) and on the laboratory acoustic emission scale (Hirata *et al.*, 1987) using a spatial two-point correlation function. It is given as follows

$$C(r) = \frac{2}{n\,(n-1)} N_r\,(R < r) \qquad (2)$$

where $N_r(R < r)$ is the number of event hypocentre or epicentre pairs with a distance smaller than r, and n is the total number of events. If the distribution of hypocentres or epicentres has a self-similar structure, $C(r)$ can be expressed in the form

$$C(r) = r^D \qquad (3)$$

where D is a kind of fractal dimension called the correlation exponent, that gives the lower limit of the Hausdorff dimensions. This method was adopted by us for investigating the fractal character of the AE hypocentre distributions of rocks at the laboratory scale (Shivakumar *et al.*, 1994; Shivakumar and Rao, 1997).

Gutenberg-Richter power-law distribution

The distribution of energy released during earthquakes has been found to obey the famous Gutenberg-Richter power law. The law is based on the empirical observation that the number N of earthquakes of size greater than m is given by the relation

$$\text{Log } N = a - bm \tag{4}$$

The precise values of a and b (scaling constants) depend on the location, but generally b is in the interval $0.8 < b < 1.5$. The above equation shows fractal behaviour of earthquakes and the b-value is strongly related to the fractal dimension of earthquakes (Aki, 1965; Hirata, 1989) in the form

$$D = 3\,b/c \tag{5}$$

In general, $c = 3/2$. Thus the above equation can be rewritten as

$$D = 2\,b \tag{6}$$

The universality of the Gutenberg-Richter relationship has in fact been observed at the laboratory scale (Hirata, 1989; Aki, 1965; Rao, 1996) and several studies are currently underway in this direction.

Generalised correlation integral function

Many natural phenomena do not have perfect homogeneous scale-invariant characteristics described by a single fractal dimension. In most cases they are bounded and heterogeneous in space. In such cases the multifractal concept, which is a natural extension of the fractal concept to heterogeneous fractals, would be more useful for analysis of the fine structure of the spatial patterns of such events as rockbursts. A multifractal is considered to be interwoven with infinitely many subfractal sets of different dimensions. Its approach involves a whole spectrum of non-integer dimensions of the set under consideration. Generalized dimensions (or multifractal dimensions) contain information on concentration clustering properties and intermittence of spatial distribution (Geilikman *et al.*, 1990). It can also be used to distinguish between a homogeneous and heterogeneous fractal set (Grassberger and Procaccia, 1983; Geilikman *et al.*, 1990; Hirata and Imoto, 1991). The generalised dimension D_q can be defined as

$$D_q = \frac{1}{(q-1)} \lim_{L \to 0} \frac{\log \sum_{i \in L} P_i^q}{\log L} \tag{7}$$

where P_i is the probability that spatial points fall into a box with size L. The parameter q can take a value ranging from $-\infty$ to $+\infty$. Among the various methods to estimate the value of D_q, the generalised correlation integral method based on the correlation integral function (Hirata and Imoto, 1991) was employed for analysing the area rockbursts and microseismic events in the Kolar gold mines (Shivakumar et al., 1996; Shivakumar and Rao, 1996). The generalised correlation integral function is given by

$$C_q(r) = \left[\frac{1}{N}\sum_{j=1}^{N}\{n_j(r)^{q-1}\}\right]^{1/(q-1)} \quad (8)$$

where N is the total number of hypocentres, and $n_j(r)$ is a local density function defined as

$$n_j(r) = \frac{1}{N-1}\sum_{j \neq k}\phi(r - |x_j - x_k|) \quad (9)$$

$\phi(s) = 1$ if $s \geq 0$
$\phi(s) = 0$ otherwise
where $|x_j - x_k|$ is the distance between two event pairs (x_j, x_k).

$$C_q(r) \propto r^{D_q} \quad (10)$$

If the spatial distribution of hypocentres of area rockbursts is fractal, then the plot of $Cq(r)$ against (r) must obey the power law, and the power-law exponent denotes its fractal dimensions. In the case of the homogeneous fractal $D_0 = D_1 = D_2 = D_\infty$, and for the heterogeneous fractal $D_0 > D_1 > D_2 > ... D_\infty$, where D_∞ is the lower limit of the fractal dimension and elucidates information about the most intensive clustering in heterogeneous fractal sets. The difference $(D_2 - D_\infty)$ represents the degree of heterogeneity and D_2 coincides with the correlation dimension.

SUMMARY OF RESULTS AND DISCUSSION

Fractal character of microcrack damage in burst-prone rocks

Correlation Dimension

A multichannel acoustic emission monitoring experiment was carried out at the GSJ, Japan on a burst-prone cylindrical amphibolite sample (100 mm height, and 50 mm diameter). The test rock was a strong (uniaxial compressive strength: 260 MPa, Young's modulus: 110 GPa) fine-grained massive amphibolite. It had a vertical joint (1.0-1.5 mm thick) with quartzite infilling. The rock was subjected to a single-stage triaxial creep under a confining pressure of 30 MPa. The differential stress was held constant at 570 MPa (approx. 85% of triaxial fracture strength of intact rock). A total number of 12,200 AE events were detected during the experiment, of which 2200 AE waveforms were recorded, and about 1800 AE events were located with an accuracy of 2-3 mm. Details of the experimental set-up, recording system and results are presented elsewhere (Shivakumar *et al.*, 1994; Shivakumar and Rao, 1997). The sequential occurrences of AE events during the various stages of creep are shown as orthographic projection plots in Fig. 1A.

The correlation dimension was calculated based on the correlation integral method (Hirata *et al.*, 1987) as discussed in the previous section. Fig. 1B shows the relationship between $C(r)$ and (r) of the jointed amphibolite during different stages of creep. The data fall on a straight line for each stage indicating that the spatial distribution of AE hypocentres has a self-similar structure throughout the scale range studied. The scale range of 3-25 mm was chosen, which was limited by the location error and sample size. The correlation dimension computed from the slopes of log $C(r)$ versus log (r) for (i) 0-108 min of creep, (ii) 108-113.8 min of creep and (iii) 113.8-114 min of creep (just prior to failure of the specimen) were found to be 0.67, 1.07 and 1.82 respectively. Since the fractal dimension characterises the degree to which the fractal fills up the surrounding space, one can predict the fracture characteristics by knowing the value of D_c. If it is 3.0, then the spatial distribution would be uniform. If it is 2.0, the spatial distribution assumes a planar structure. If it is 1.0, it represents a linear structure in 3-D space. In the present experiment during the initial stages of creep, the fractal dimension was low and tended to increase with progress of creep. This is because of the concentrated distribution of AE, forming denser clusters around a pre-existing flaw (Fig. 1A) during the initial stages of creep, and diffused distribution of AE along the eventual fracture plane during subsequent creep stages (Fig. 1A). These observations indicate that microcracks which concentrated more on the joint plane during the incremental loading and primary creep (Fig. 1A) weakened the material, resulting in low D_c. Subsequently, the microcracking activity during secondary and tertiary creep regimes shifted onto the eventual fracture plane (Fig. 1A). The spatial distribution of AE however was diffused, resulting in a higher D_c (Fig. 1B). On the other hand, when the material softens with no migration of microcrack clusters, the softening process is accompanied by shear band formation with dense clusters of AE, which results in a lower correlation dimension, as observed in Oshima granites (Hirata *et al.*, 1987).

Seismic b-value

Among the various AE parameters, amplitude distribution and seismic *b*-value have proven most useful. The power-law exponent, D, inferred from the seismic *b*-value, measures the relative proportion of large and small cracks (Aki, 1965). However, determination of seismic *b*-value by

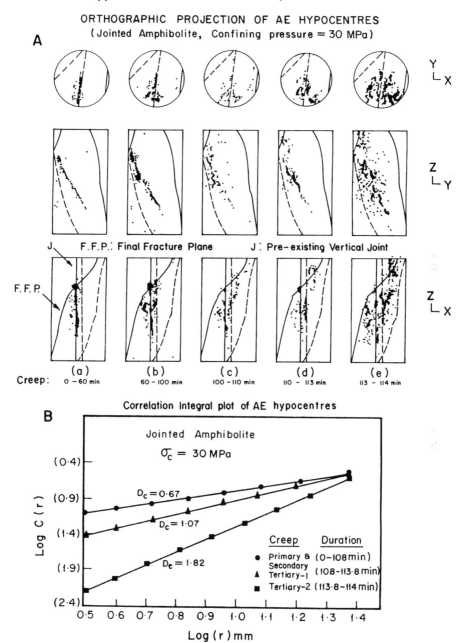

Fig. 1A Orthographic projection showing the spatial distribution of AE hypocentres in jointed amphibolite during various stages of creep under triaxial compression

Fig. 1B Correlation integral plot from the distribution of AE hypocentres of jointed amphibolite under creep. The values of D_c at various stages of creep are also shown in the plot

using commercially available AE systems is not straightforward owing to the masking effect suffered by a large number of weak AE events during the inelastic deformation of the test rock. This can be overcome by processing the individual subsets of AE peak amplitudes in sequential non-overlapping time and stress windows (Rao, 1996). This yielded useful results not only of the cumulative exceedence frequency plots, but also the average amplitude of AE for each of the stress intervals chosen. From this, an inference can be drawn about the average crack length (i.e., average amplitude, $\overline{A}_i$) in any stress interval. Further, the b-value and D can be computed using the following equations (Aki, 1965) at any desired stress interval:

$$b = \frac{20 \log_{10} e}{(\overline{A}_i - A_0)} \tag{11}$$

$$D = 2b \tag{12}$$

The results obtained as above for the burst-prone rocks of the Kolar gold mines are shown plotted as a function of failure stress (Fig. 2). Both the burst-prone rocks show a universally acceptable b-value of around 1.0 (i.e., $D = 2$) until dilatancy commences at stresses > 80 % of failure stress. As failure approaches, the rocks generate relatively high amplitude events accompanying the formation and growth of longer cracks. Consequently, a sharp fall in b-value (or D) in rocks such as dolerite (Fig. 2) and with fluctuations in schistose rocks (Fig. 2) to reach a minimum of about 0.5 ($D = 1$) is observed in accordance with the physical processes underlying crack growth in deforming rocks. Thus in heterogeneous materials such as rocks, the evolution of damage towards the critical condition, $D = 1.0$, corresponds to a transition from stable damage development by many microcracks to the unstable crack growth of crack coalescence, or a single macrocrack.

Fractal character of area rockbursts

The qualifying data were short-listed from the data bank for investigating the fractal character of the spatial distribution of the hypocentres of three major area rock bursts (ARB) of the Champion reef mine of KGF (Shivakumar *et al.*, 1996). The ARB-I was at a shallow depth (approx. 2000 ft) and not affected by major geological features, whereas ARB-II was quite deep (about 9000 ft) and also influenced by faults, folds and pegmatite intrusions (Fig. 3). The ARB-III was at an intermediate depth (approx. 4000 ft). The ARB-I consisted of a series of rockbursts occurring in abandoned and excavated regions of the Champion reef mine. During the first hour after the major rockburst of local magnitude of 2.48, 38

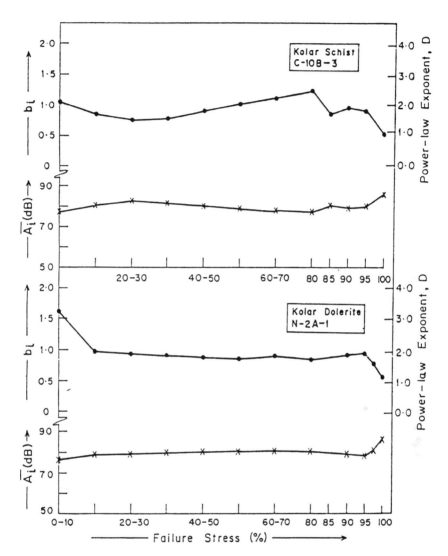

Fig. 2   Plot showing the average amplitude, seismic *b*-value and power-law exponent *D* of the population of AE in burst-prone rocks as a function of applied stress. The rocks were stressed to failure under uniaxial compression

bursts were found located in the same area. During the next 14 days, as many as 125 rockbursts covering a strike length of approximately 2500 ft or 750 m took place in the same region (Fig. 3). The ARB-II consisted of a series of rockbursts numbering 64 of minor to medium intensity, that occurred after the major rock burst of local magnitude 2.44 over a period of one week in region II (Fig. 3). Most of the events of ARB-II had their

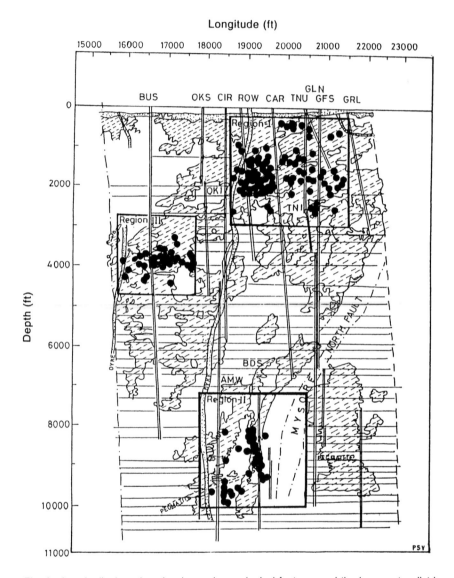

Fig. 3 Longitudinal section showing major geological features and the hypocentre distribution of area rockbursts I, II and III of Champion reef mine. Shafts are represented by two parallel lines. Excavated area is indicated by hatching

foci clustered along the current stoping regions. The ARB-III also consisted of a series of rockbursts (66 events) of major to medium intensity
followed by a major rockburst of local magnitude 3.09 in region III
(Fig. 3). These events occurred over a period of 15 days in the old workings. The time gap between ARB-I and ARB-II was nearly 3 years, while
ARB-III followed ARB-II after only 20 days.

Multifractal Structure of Area Rockbursts

Generalised dimensions, D_q, of the spatial distribution of area rockbursts were calculated from the slopes of the generalised correlation integrals, $C_q(r)$, versus distance r, on a log-log plot by the least squares method. The positive values of q, which range from 2 to 10 (Figs. 4-6) were made use of. Correlation integral functions showed that the spatial distribution of hypocentres of all the three area rockbursts have a multifractal structure. With increase in q, the change in D_q reduces and when $q > 10$, D_q changes slightly. Hence, it is reasonable to use D_{10} as an approximation for D_∞. The self-similarity of hypocentres of all the three area rockbursts is apparently valid for the scale range studied (200-1200 ft). The spatial distribution of all the three area rockbursts indicates that they are heterogeneous with generalised dimension $D_2 = 2.10$; 1.58 and 1.95 for ARB-I, II and III respectively. The degree of heterogeneity (i.e., $D_2 - D_\infty$) was found to be 0.52, 0.37 and 0.41 respectively for the three area rockbursts. Among the three ARBs, the multifractal dimensions of ARB-II are relatively low because it occurred in a highly stressed region containing prominent geological features such as the Mysore north fault, pegmatite inclusion and folded nature of the lode. The near-by working stopes have enhanced the stress in that region to a more critical state.

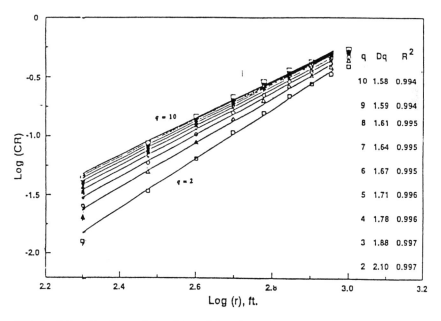

Fig. 4 Generalised correlation integral functions of hypocentre distribution of area rockbursts-I. Fractal dimensions for $q = 2-10$ are specified on the right side along with the correlation coefficient (R^2)

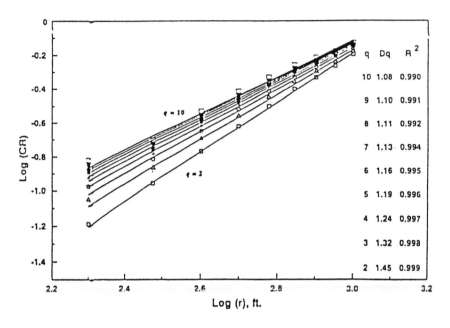

Fig. 5 Generalised correlation integral functions of hypocentre distribution of area rockbursts-II. Fractal dimensions for q = 2-10 are specified on the right side along with the correlation coefficient (R^2)

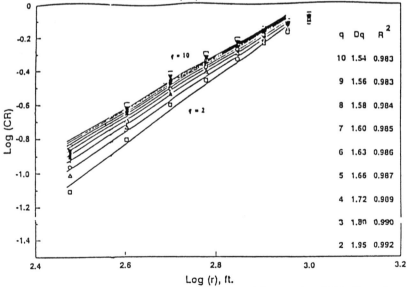

Fig. 6 Generalised correlation integral functions of hypocentre distribution of area rockbursts-III. Fractal dimensions for q = 2-10 are specified on the right side along with the correlation coefficient (R^2)

Fractal character of microseismic events

Analysis of distribution of foci of microseismic events with respect to mine workings is very important because it serves as a precursor to rockburst. Two clusters of microseismic events (MS-I and MS-II) associated with the Glen ore shoot region of the Champion reef mine (Fig. 7) were analysed from the fractal point of view in both space and time. The MS-I was associated with the major rockburst of 1989 at level 98, while

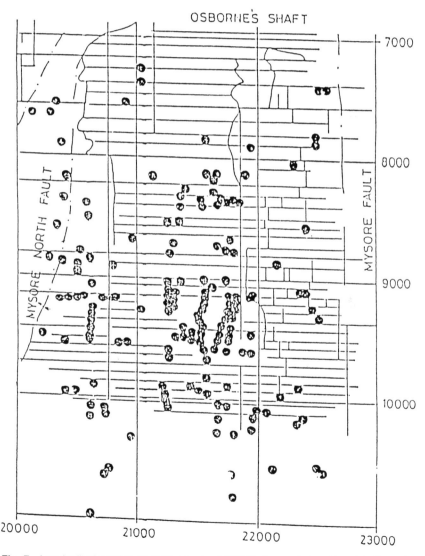

Fig. 7   Longitudinal section of Glen ore shoot (FOCI of recorded rockbursts) Champion reef mine

MS-II was associated with the major rockburst of 1990 at level 103. The multifractal technique described above was used for investigating the changes in correlation dimension in space as well as time. In the case of time fractals, time intervals instead of pair distances were used (Coughlin and Kranz, 1991) similar to the two-point correlation method described for space fractals.

Figures 8 and 9 show the multifractal structure of microseismic events. The MS-I (Fig. 8) has relatively high correlation dimension, D_2 with higher degree of heterogeneity ($D_2 - D_\infty = 0.73$) compared to MS-II (Fig. 9). The MS-II exhibits a band-limited fractal (bifractal). This could be a result of different mechanisms operating perhaps at different scales. On close examination we found that the recorded events of MS-II contained many blast-generated events, which might have a different spatial distribution. Probably this intermix of events resulted in a bifractal structure. The spatial patterns of MS-II indicate higher stress concentration ($D_2 = 1.79$)

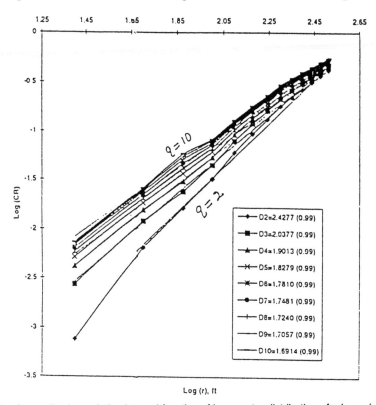

Fig. 8   Generalised correlation integral function of hypocentre distribution of microseismic events. Fractal dimensions for $q = 2$-10 are specified on the right side along with the correlation coefficient

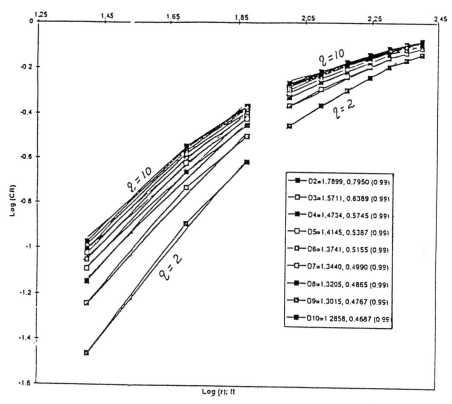

Fig. 9 Generalised correlation integral function of hypocentre distribution of microseismic events. The band-limited fractal dimensions for $q = 2\text{-}10$ are specified on the right side along with the correlation coefficient

and smaller heterogeneity ($D_2 - D_\infty = 0.51$) compared to those associated with MS-I. These differences arise due to the stress and strength heterogeneity variations of the rock and the depth at which the bursts took place.

When the two-point correlation method was applied to the time intervals instead of hypocentre distances, the power-law distribution of time intervals was found to hold good for several hours (Figs. 10 and 11), and different slopes are due to the different distribution of time intervals between the events. However, interestingly the spatial and time clustering of MS-I and MS-II events did not show one-to-one correspondence. That is, events of MS-II show a higher degree of clustering in space and a lower degree of clustering in time compared to the events of MS-I. Therefore, it is necessary to verify whether or not the decrease in D of both spatial and temporal clustering instead of spatial clustering alone of microseismic events should be taken as a more positive precursor for predicting and controlling rockbursts.

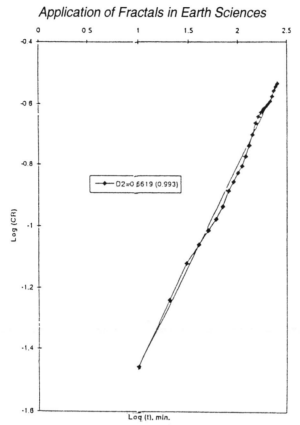

Fig. 10 Generalised correlation integral function of microseismic events (time) distribution. The correlation dimension is specified on the left side along with the correlation coefficient.

CONCLUSION

The potential of fractal techniques as applied to acoustic emission, seismic and microseismic events to quantify and describe the fractal character of fracturing process in rock at different scales has been demonstrated. Laboratory experiments revealed several interesting features of microcrack damage and the influence of major discontinuities in terms of the formation, growth and migration of AE clusters in the burst-prone rocks of Kolar gold mines. It was confirmed that the fractal character of rock fracture and rockburst is mainly controlled by the state of stress distribution and heterogeneity of rock media. Sometimes a single fractal dimension may not suffice to describe the spatial distribution of seismicity. Multifractal analysis would be more useful in characterising such events. The change in fractal character of different area rockbursts indicates that they are bounded by different heterogeneous stress fields associated with different mining and geological conditions.

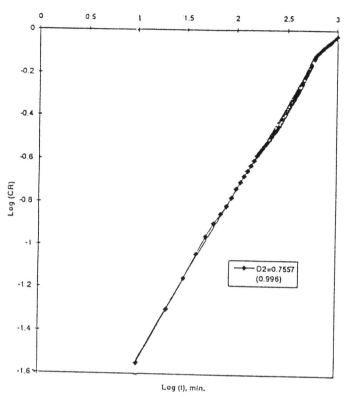

Fig. 11 Generalised correlation integral function of events (time) distribution of microseismic events. The correlation dimension is specified on the right side along with the correlation coefficient

ACKNOWLEDGEMENTS

Experimental studies at the laboratory scale and in the underground gold mines at KGF became possible with the support of an Indo-Japan collaboration project between NIRM (India) and GSJ (Japan) while Dr. Rao was working at the NIRM, KGF during 1992-95. This paper is released for presentation and publication with the kind permisssion of Dr. N.M. Raju, Director, NIRM and Dr. H.K. Gupta, Director, NGRI.

REFERENCES

Aki K. 1965. Maximum likelihood estimate of b in the formula Log $N = a - bm$. *Bull. Earthq. res. Inst.*, **43**: 237-239.

Coughlin J and Kranz R. 1991. New approaches to studying rockburst-associated seismicity. *Proc. 32nd U.S. Rock Mech. Symp.* Balkema Publ., USA, pp. 491-500.

Geilikman MB, Golubeva TV and Pisarenko VF. 1990. Multifractal patterns of seismicity. *Earth Planet. Sci. Lett.*, **99**: 127-132.

Grassberger P and Procaccia L. 1983. Measuring the strangeness of strange attractors. *Physica.*, **9D**: 189-208.

Hirata T. 1989. A correlation between the *b*-value and the fractal dimension of earthquakes. *J. Geophys. Res.*, **94**: 7507-7514.

Hirata T, Satoh T and Ito K. 1987. Fractal structure of spatial distribution of microfracturing in rock. *Geophys. J.R. Astron. Soc.*, **90**: 369-374.

Hirata T and Imoto M. 1991. Multifractal analysis of spatial distribution of micro-earthquakes in the Kanto region. *Geophys. J. Int.*, **107**: 155-162.

Kagan YY and Knopoff L. 1980. Spatial distribution of earthquakes: The two-point correlation function. *Geophys. J.R. Astron. Soc.*, **62**: 303-320.

Lei X, Nishizawa O, Kusunose K, and Satoh T. 1992. Fractal structure of the hypocenter distributions and focal mechanism solutions of acoustic emission in two granites of different grain sizes. *J. Phys. Earth*, **40**: 617-634.

Mandelbrot BB. 1982. *The Fractal Geometry of Nature*. Freeman Publ., San Francisco.

Rao MVMS. 1996. Significance of AE-based *b*-value in the study of progressive failure of brittle rock: Some examples from recent experiments. *Proc. 14th World. Conf. on Non-Destructive Testing.*, vol. 4, pp. 2463-2467. New Delhi, India.

Rao MVMS, Kususnose K, Shivakumar K and Srinivasan C. 1994. Microseismic and acoustic emission monitoring experiments in Nundydroog mine. Tech. Rep. IJCP/93-96/I of the India-Japan S & T Collaboration Project., pp. 1-27. Report submitted to DST (India) and MITI (Japan), Nov. 1994.

Shivakumar K and Rao MVMS. 1997. Fractal analysis of acoustic emission and its application to the study of rock fracture and rockbursts. *Proc. 4th Natl. Workshop on Acoustic Emission*, pp. 80-102. Mumbai, ISNT Publ.

Shivakumar K, Rao MVMS, Srinivasan C and Kusunose K. 1996. Multifractal analysis of spatial distribution of area rockbursts at Kolar gold mines. *Int. J. Rock Mech. Min. Sci., & Geomech. Abstr.*, **33**: 167-172.

Shivakumar K, Satoh T, Nishizawa O, Kusunose K and Rao MVMS. 1994. Micromechanics of fracturing of jointed amphibolite as inferred from spatio-temporal distribution of AE hypocenters and surface strain map. Interim Report # 2, Indo-Japan S & T Collaboration Project/93-96, 1994.

Srinivasan C, Arora SK and Yaji RK. 1997. Use of mining and seismological parameters as premonitors of rockbursts. *Int. J. Rock Mech. Min. Sci., & Geomech. Abstr.*, **34**: 1001-1008.

Xie H. and Pariseau WG. 1993. Fractal character and mechanism of rockbursts. *Int. J. Rock Mech. Min. Sci., & Geomech. Abstr.*, **30**: 343-350.

FRACTAL DIMENSION ANALYSIS OF SOIL FOR FLOW STUDIES

V.P. Dimri*

INTRODUCTION

Groundwater and subsurface soil are often contaminated by toxic substances such as pesticides. Groundwater is the precious source of our drinking water and is also needed in agriculture and horticulture. The huge amount of fertilisers and pesticides might be necessary for increasing the products and protecting the plants from hazardous insects. But fertilisers are often left in the soil if they are not fully utilised by the roots of the plants. The remaining fertilisers and pesticides then get mixed with water and infiltrate the soil.

The flow of water in soil follows a very complex route. There are many factors which affect the flow. These are mainly lithology, texture of soil and soil moisture. The flow also depends on complex hydrological, mineralogical and chemical heterogeneity of the medium. Recent studies (Hagrey *et al.*, 1997) have shown that water flows in the vadose zone before it reaches the aquifer in a preferential manner, such as 'fingering' and or macropores.

The word 'fingering' was coined for the instability condition wherein water drives oil out of porous media (Feder, 1988). The concept of viscous fingering has been described by Srivastava (1999). The criteria for

* National Geophysical Research Institute, Hyderabad, 500 007, India.

fingering is that the viscosity of fluid driving in is smaller than the viscosity of fluid driven out. Studies have shown that the fingering in porous media is fractal. It definitely depends on pore geometry. Also, the study of percolation by finding the cluster of pores and percolation clusters which allow the fluid to flow from one end to another in the porous media is important. The irregular geometry of a pore structure can be studied by the fractal theory.

FRACTAL PORES

In this paper we investigate the pore structure of subsurface soil using fractals since many experimental works have shown that pore surfaces are fractal (Thompson, 1991; Feder, 1988). Many authors have investigated the fractal dimension of various soils from many parts of the world and indicated that most of the soils are fractal with the exception of novaculite, found to be non-fractal (Kruhl, 1988).

The purpose of this paper is to demonstrate the method of finding the fractal dimension of a soil. Recently, I saw a model experiment carried out at the Institute of Geophysics, University of Kiel, Germany. It was designed to see the fingering structure in a soil brought from an agricultural field about 30 km from Kiel. A large-scale tank 5 m × 3 m × 2 m filled with sand was constructed in the premises of Kiel University to study in detail the preferential flow using radar and resistivity measurements (Hagrey *et al.*, 1977).

Sands differ in type, from fine silt to coarse sand of varying size. Table 1 shows the grain size (average diameter) and average cumulative mass following the standard density of various soils.

Table 1 Different types of sand used in the tank experiment (Hagrey *et al.*, 1977)

Class	Size	Range (dia.)	Aver.	Sample 1	Sample 2	Aver.	Cum.
Silt	Fine	2.0-6.3	4.15	0.00	0.05	0.025	0.025
	Medium	6.3-20.0	13.15	0.10	0.15	0.125	0.150
	Large	20.0-63.0	41.50	1.70	0.95	1.325	1.475
Sand	Fine	63-200	131.50	32.40	32.25	32.000	33
	Medium	200-630	415.00	65.50	68.35	66.000	100
	Large	630-2000	1315.00	1.15	1.25	1.200	101

From the fractal theory a relation between the number of grain particles and their radius is given (see e.g., Turcotte, 1986) as

$$N(r<R) \propto R^{-D} \qquad (1)$$

where $N(r<R)$ is the number of particles of radius R or greater and D is fractal dimension.

Since counting the number of particles is not possible, this has been replaced by volume or mass, a more readily measured quantity. So the above equation becomes

$$\frac{M(r < R)}{M_t} = \left(\frac{R}{R_L} \right)^{3-D}$$

(2)

where M represents the soil mass and M_t the total mass.

From the above equation the fractal dimension can be estimated. We have plotted the grain size (radius) and average cumulative mass of sand in a log-log plot (Fig. 1). A straight line using the least squares method was fitted to satisfy all six points. The slope of the straight line and, using the above equation, the fractal dimension of the sand were obtained. The latter is 2.4832 with a correlation coefficient of 0.9371.

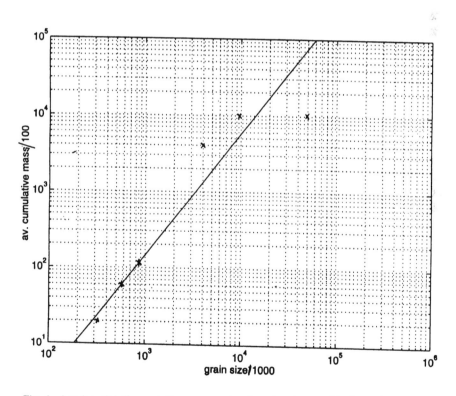

Fig. 1 Log-log plot of cumulative mass and radius of grains used in the tank experiment

The fractal dimension of several soils was studied earlier by various authors (Turcotte, 1986; Kruhl, 1988; Tyler and Wheatcraft, 1990 and their other papers). It was found that coarse grains have a smaller fractal dimension than fine grains such as clay, whose fractal dimension may approach 3. Fractal dimensions change with the content of sand and/or clay present in the soil and results of various investigations can be summarised as below:

Soil	D
Sand	2.41
Loamy sand	2.56
Sand clay loam	2.76
Sand clay	2.86
Clay	2.96-3.00

Thus the soil under discussion can be classified as loamy sand. The medium is fractal. According to a field study in Sweden, herbicides are transported quite deep into loamy sand. They further observed that the bulk mass of herbicides (>90%) nonetheless remains in the top layer of such loamy sand.

It is interesting to note that the last three points of Fig. 1 can be fitted by another straight line. This may indicate a bimodal fractal system having structure and textural fractal dimension (Sukhtankar, 1999) which may be part of a future study.

The value of fractal dimension can be used to estimate the porosity of soil (Kartz and Thompson, 1985), an important factor for the transport of flow. Porosity has been predicted from the fractal dimension estimated from grain size distribution as shown here. So, the fractal study of soil would help us to understand the mechanism of onset of flow in the fractal media.

CONCLUSION

Soil media can be fractal at different scales. A medium can be homogeneous and can be fractal. The flow in homogeneous media has been studied extensively but there is need to study flow in fractal media. A study in this direction is already underway, for example see Alder (1996) and other references therein.

ACKNOWLEDGEMENT

I am grateful to Dr. H.K. Gupta, Director NGRI for permission to publish this work. This part of the work was carried out when I was at the Institute of Geophysics, University of Kiel, Germany with a senior DAAD/CSIR fellowship. I thank Prof. R. Meissner for suggesting this problem.

REFERENCES

Alder PM. 1996. Transport in fractal porous media. *J. Hydrology*, **187**: 195-214.

Feder J. 1988. *Fractals*. Plenum Press, NY

Hagrey SA, Klempnauer S, Michaesen T and Kanemann. 1997. *Preferencial Flow*. Poster presented at 57th Annual Meeting, German Geophysical Society (DGG), Potsdam.

Katz AJ and Thompson AH. 1985. Fractal sandstone pores: implications for conductivity and pore formation. *Phys. Rev. Lett.*, **54**: 1325-1328.

Kruhl CE. 1988. Fractal measurement of sandstones, shales and carbonates. *J. Geophys. Res.* **93**: 3286-3305.

Srivastava V. 1999. The Percolating Fractals. *Application of Fractals in Earth Sciences* (this volume), Dimri, V.P. (ed.)

Sukhtankar RK. 1999. Fractal, fractal dimension and geology. *Application of Fractals in Earth Sciences* (this volume), Dimri, V.P. (ed.).

Thompson AH. 1991. Fractals in rock physics. *Ann. Rev. Earth Planet Sci.*, **19**: 237-243.

Turcotte DL. 1986. Fractals and fragmentation. *J. Geophys. Res.* **91**: 1921-1926.

Tyler SW and Wheatcraft SW. 1990, Fractal processes in soil water retention. *Water Resource Res.*, **26**: 1047-1054.

DETECTING CHAOS FROM GEOPHYSICAL TIME SERIES

INTRODUCTION

Prediction of a physical phenomenon is an interesting curiosity and the
ultimate goal of science. Coupled terrestrial and extraterrestrial dynamics follow certain laws of physics and exhibit intrinsic order and har-
mony. However, a famous mathematician, H. Poincare, of the late nine-
teenth century discovered that certain evolving mechanical systems governed by Hamilton's equations could display chaotic motions. Later, in
1963, the noted meteorologist, E.N. Lorenz provided a more definite
evidence of chaos; he showed that even a simple set of deterministic first-
order coupled non-linear differential equations can lead to chaotic trajectories. Most recently, Sussman and Wisdom (1988) have shown quantitative evidence that motion of the planet Pluto is chaotic. Evidence of chaos
has now been discovered in many geophysical geo-bio-ocean-atmospheric
systems, geological records and extra-terrestrial system. What do we mean
here by 'Chaos'? It might describe a notion of disorder/irregularity and
unpredictability. 'Chaos' possibly means that a physical phenomenon
that may appear to be fluctuating randomly at first glance upon closer
examination may be found to possess considerable regularity. The dynamical system may apparently exhibit certain intrinsic aperiodic com-

---

* Theoretical Geophysics Group, National Geophysical Research Institute, Hyderabad
500 007, India.

ponents with some arbitrary components. The obsevations and model studies indicate that length of day (LOD) system dynamics may combine both stochastic and some regular components, in which the apparent regula-rity might be related to the inherent non-linear dynamics in the interacting system, but the arbitrariness probably stems from the sensitive dependence on initial conditions. The chaos thus adds a new quality of irregular stochasticity with some degree of freedom (Kaiser, 1990).

In a chaotic system, similar patterns reappear rendering a deceptively periodic look but sequences do not repeat exactly. This is due to the fact that weak or strong non-linear linkages among the state variables cause small errors in measurements of the initial conditions. This error grows exponentially and evolving nearby trajectories diverge and eventually become uncorrelated (Vassiliadis *et al.*, 1990). In order to make a precise statement, or quantify predictive time limit, it is essential to examine the behaviour of various non-linear parameters, e.g. strange attractor, Lyapunov exponent, K_2 entropy, spectral characteristics and complex phase-space trajectories etc. in composite chaotic time signals. An appropriate analysis of coupled dynamical time series and identification of various properties of chaotic attractor render a way of precise forecast of the state of the dynamical system.

LIMIT CYCLE, TORUS, STRANGE ATTRACTOR AND PHASE-SPACE TRAJECTORIES

The various evolutionary fluctuations exhibit the features of a dissipative dynamical system with possible attractors in phase space. Chaotic or non-chaotic system (or in other words random and/or deterministic components of a physical process) can be classified using the embedding theory of a non-linear dynamical system (Takens, 1981). If all the trajectories of evolutionary processes converge towards a 'subset' of phase space, irrespective of their initial conditions, the subset or 'submanifold' is called an 'attractor' (Tsonis and Elsner, 1988). An interesting discussion on chaos physics and attractors is given by Lauterborn and Parlitz (1988). Accordingly, the nature of attractor is given by its fractal dimension. For example, a physical process possessing a point attractor or limit cycle has periodic trajectories as its attractors and its dimension are unity like the oscillation of a pendulum. A quasi-periodic process exhibiting several non-harmonic frequencies may have a torus as its attractor. This may happen in a physical situation in which the system oscillates with two or more incommensurable frequencies. For instance, due to inherent non-linearity the earth's precessional frequency split in a 23,000-year and

19,000-year period, which is also imprinted apparently in Pleistocene climate indicator records. The splitting of the main signal possibly arises due to inherent non-linearity, the essential and chief cause of chaos. In contrast, a 'strange attractor' will have a clearly non-integer dimension (fractal dimension) and can be represented by a more complex form in a phase space. It is characterised by chaotic trajectories and a broadening of the frequency spectrum. Evidence of chaos/strange attractor has been reported in many global geophysical and planetary system dynamics. To make the definitions of different attractors, e.g., limit cycle, quasi-periodic attractor and strange attractor more explicit for an uninitialized reader Fig. 1(a, b, c, d) displays four different kinds of attractors (e.g. point attractor, periodic limit cycle, quasi-periodic and chaotic attractors, respectively. The fractal dimension of a chaotic system (strange attractor) indicates the number of ordinary differential equations for modelling the process, or in other words, is indicative of the number of variables required for almost complete description of the dynamic system (Tsonis and Elsner, 1988).

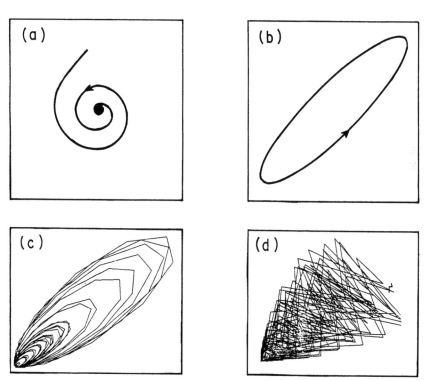

Fig. 1 Types of attractors

NATURE OF PHASE-SPACE TRAJECTORIES AND SPECTRAL SIGNATURES OF CHAOTIC TIME SERIES

LOD data

As an illustrative example of the nature of phase-space trajectories, power spectra and other non-linear parameters, we have worked with an excellent quality of daily variations of Length of Day (LOD) data for the last 30 years (1962-1992). The data have been obtained from high precision space-based geodetic techniques such as Satellite Laser Ranging (SLR) and Very Long Base-line Interferometry (VLBI). The data set exhibits minute but apparently complicated long- and short-term changes in LOD, measurable up to several parts in 10^8. These changes manifest a variation of up to several milliseconds in the Earth's axial rotation, possibly caused by nonlinear interaction among various oceanic and atmospheric processes. The LOD time series available here comprises 10,580 data values spanning the period of 1962-1992. The LOD data set is the time derivative of UT-AT, where UT is universal time and AT (Atomic reference time, 86,400 s). The original (UT-AT) data were obtained from the Bureau International de l'Heure (BIH) (M. Feissel, pers. comm.). The apparently quasi-periodic LOD variations, displayed in Fig. 1, can be classified into three spectral bands: (i) the longer period components of secular changes, which are produced by tidal friction (Sun-Moon system) and from internal sources such as changes in moment of inertia of solid Earth; (ii) irregular 'decadal variations' (up to several milliseconds/century in the LOD) are caused possibly due to angular momentum transfer between the Earth's solid mantle and fluid core; and (iii) the unpredictable and most rapid 'non-tidal' variations on time scales ranging from days up to a few years. These are largely of atmospheric origin (up to about 1 millisecond in amplitude).

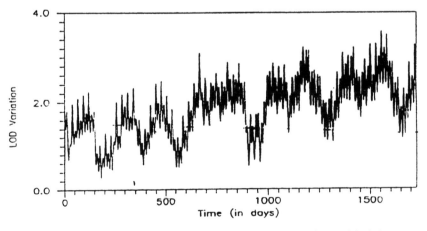

Fig. 2   A representative portion of LOD time series drawn from original data

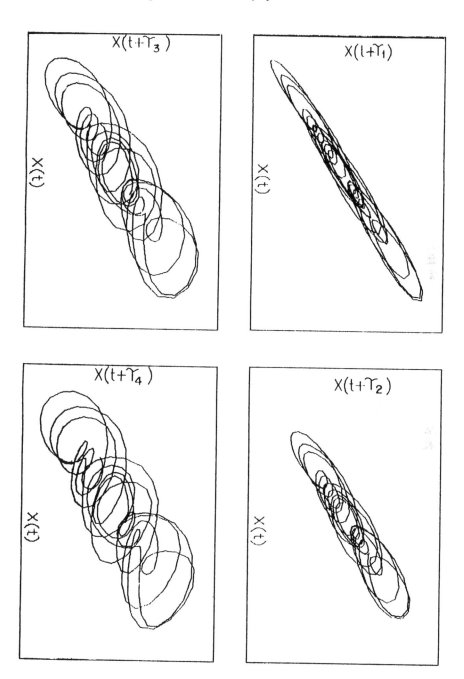

Fig. 3 Two-dimensional phase-space trajectories (*x, x + y*) of LOD time series

Phase-space trajectories

Figure 3 displays the nature of trajectories in the two-dimensional $(XX+\tau)$ phase space. The projection of $X\ X + \tau$ in the two-dimensional phase plane provides qualitative visualisation of the rate of change of folding and stretching of LOD/AAM signal. The associated stretching and folding transform the shape of the error from normal to non-uniform, creating fractal shape distribution, and eventually destroy all information regarding initial conditions. Apparently it exhibits a somewhat complex orbit (periodic, quasi-periodic and irregular). Even in a weakly non-linear system, the same orbit displaces the system state from a period either due to intrinsic chaotic dynamics or the interactions of frequencies from different modes (Ott and Spano, 1995). Figs. 3 and 4 depict two- and three-dimensional phase trajectories respectively. From Figs. 3 and 4, one can visualise a somewhat complex and unstable LOD/AAM structure, indicating chaotic evolution.

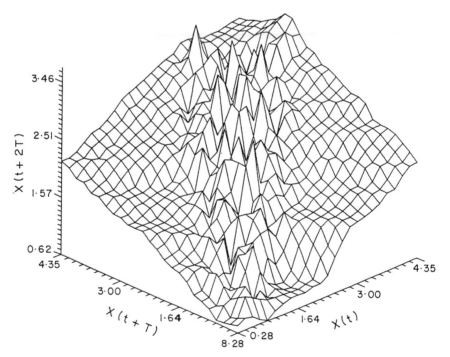

Fig. 4 A three-dimensional phase-space presentation of LOD time series

Power spectra

The Fourier spectrum of LOD data shows broadbanded and noisy spectral patterns (Fig. 5). There is evidence of several tidal components with

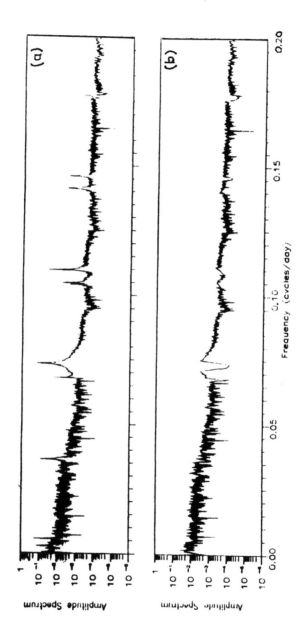

Fig. 5 Power spectra: (a) original LOD data, (b) filtered LOD data

dominant spikes clustered around 27, 13, 9 and 7 days etc. and non-random seasonal components at 372 and 182 days (Fig. 5a). The common practice to perform non-linear analysis is to properly filter out these periodic tidal signals from the data series and then study the coupled non-linear dynamics. Accordingly, elimination of all dominant periodic tidal signals are performed by using the method of Delache and Scherrer (1983). This is carried out by using the fast Fourier transform (FFT) method for 2^{13} data series. After obtaining FFT of the detrended original LOD time series, the Fourier coefficients of the dominant periodic are successively set to zero and an inverse transform taken. Thus the resultant time series is deprived of all periodic components (Fig. 5b).

ANALYSIS OF ATTRACTOR DIMENSION, KOLMOGOROV ENTROPY AND LYAPUNOV EXPONENT USING A DISCRETE TIME SERIES

Attractor dimension

The mathematical basis of chaos is fairly well understood. Computational procedures for attractor dimension follow the embedding theory of Takens (1981). Our emphasis here is on an accurate and reliable reconstruction of the system's attractor from detailed time series. Here, a given scalar time series $X(t_i)$, $i = 1, 2, ..., N$ is embedded in an m-dimensional phase space using appropriate delay times. The vector x_i can be constructed in m-dimensional phase space ($m = 1, 2, 3,...$) from the data sequence $X(t_i)$ as follows:

$$X_i = [X(t_i), X(t_i + \tau), ... X(t_i + (m - 1)\tau)] \tag{1}$$

$$i = 1, 2,..., N$$
$$m = 1, 2,..., n.$$

Here m is the 'embedding dimension' and τ is time delay. A suitable choice of the time lag τ is quite important in the attractor dimension analysis. In practice, the first minimum from the auto-correlation function analysis of the time series under study is preferred (Roberts, 1991). Evidently for the m dimensional phase space, $m - 1$ additional time series are required. The time series is obtained from the original series and by a time shift of $(m - 1)t$. The m dimensional phase-space trajectory will have the m time series as its components. Computation of correlation function $C_m(r)$ proceeds using the scheme of Grassberger and Procaccia (1983):

$$C_m(r) = \lim \frac{1}{N^2} \sum_{i=1}^{N} \sum_{j=1}^{N} \theta \left(r - \left\| X_i - X_j \right\| \right) \tag{2}$$
$$i \neq j$$

where θ is the Heaviside step function which satisfies the condition $\theta(X) = 0$ for $X < 0$ and $\theta(X) = 1$ for $X > 0$, N is the size of the data set. For a suitable value of r as $r \rightarrow 0$, $C_m(r)$ follows the power law and is given by $C_m(r) \sim r^d$, where d is correlation dimension of the attractor (Fig. 6)

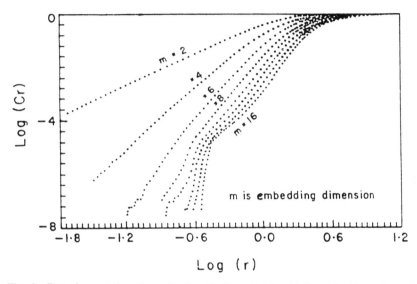

Fig. 6 Plot of correlation dimension log $C_m(r)$ against log (r) for embedding dimension 2-16

Here, we have sufficient number of data points. Early studies have shown that to obtain a reliable estimate of the attractor dimension, the number of points required is of the order of 10^d, where d is attractor dimension. Some workers have demonstrated that for a data set of known complexity, the existence of a scaling region is critically dependent on the amount of data (Tsonis and Elsner, 1988). Recent theoretical studies have indicated, however, that the number of data points required may be reduced, approximately, to the order of $10^{2 + 0.4d}$ (Triantafylloy *et al.*, 1995). In some cases, it has been shown that even 500 data points can reliably extract the attractor dimension (Abraham *et al.*, 1986). These theoretical uncertainties require certain suitable procedures and checks to be followed to estimate the attractor dimension. Analysis of LOD data using the algorithm discussed above indicates a low dimensional fractal dimension of the order of 5-6 in LOD time series.

Low dimensionality (limited degrees of freedom) provides first-order information about the complexity of the LOD system dynamics. Fractal dimension indicates the minimum number of variables present in the evolution of the LOD system (Tsonis and Elsner, 1988). To discern the

exact nature of the system's evolution it is essential to have a deep understanding of the physical mechanisms and their non-linear interactions
with many other coupled interactive systems and recognition of a limited
number of well-defined parameters which are likely to be involved in the
evolution dynamics.

Comparison with Random Walk Model

It would be prudent to make a comparison of these results with random
phase-space generated by a non-linear stochastic model. A first-order
autoregression model of the form:

$$y_t = e^{y_{t-1}} + \varepsilon_t, \qquad t = 1, 2\ldots$$

can be used where ε_t are the normal independent random variables uniformly distributed in the interval 0 to 1. The maximum likelihood estima-
tor is calculated from the LOD data (Fuller, 1976). We have repeated
similar analysis using exactly the same number of randomised data points.
Figure 7 comperes the results of original, filtered (tidal frequencies)

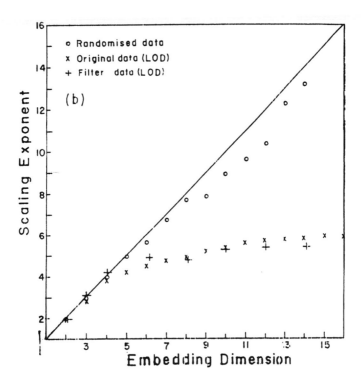

Fig. 7 Convergence of slope of original LOD and filtered LOD data against embedding
dimension. The plot of white noise process and slope calculated from randomising
the LOD data is also shown

and random data. The convergence of slope of filtered data differs slightly from the original data, but obviously indicates the exis-tence of a deterministic component. It is clearly evident, however, that estimated dimension after randomisation of the original data closely follows the white noise process.

Calculation of Lyapunov exponent

The time evolution $x(t)$ of a non-linear system can be defined by a non-linear differential or recurrence deterministic equation of the form:

$$dx(t)/dt = X[x(t)] \qquad (3)$$

for continuous time t, or

$$x(t + 1) = f[x(t)] \qquad (4)$$

for discrete time. This is also called an 'iterative map'.

Here $x(t)$ is a finite or infinite dimensional vector and f is some non-linear function of x. In the case of a discrete time series, eqn. (4) for a given initial condition, can be used to generate the state of the dynamical system from the present state. The variable defines a state space in which the system is represented at each instant by a point and traces out a trajectory or orbit during its time evolution. Stability of the orbit can be quantified by the Lyapunov exponent. The Lyapunov exponent is a measure of chaos of orbital instability. A chaotic process is highly dependent on the initial conditions which implies that two processes which are governed by the same equations and start from two slightly different initial states x and $x + \Delta t$ diverge after a time t at an exponential rate (Fig. 8). Following Ruelle (1994), if $\delta X(0)$ is an infinitesimal change of the initial condition for the LOD/AAM system and $X(t)$ is the corresponding change at a later time $t > 0$, (Fig. 8), then in general:

$$\delta X_i(t) = \exp(\lambda t) \mid \delta x_i(0) \mid \qquad (5)$$

More precisely, if $X(t)/X(0)$ denotes the norm of the Jacobian matrix, there is a limit

$$\lambda = \lim_{t \to \infty} (1/t) \sum \log \left\| \delta X(t)/\delta X(0) \right\| \qquad (6)$$

which is called the largest Lyapunov exponent. The above equation states that λ is the rate of separation (per unit time) of orbit passing near $X(0)$ (Tsonis and Elsner, 1988). If $\lambda > 0$, the orbit diverges exponentially and the system is chaotic and long-term predictability is lost. When no positive exponent exists, there is no exponential divergence and thus long-term predictability is guaranteed (Vassiliadis *et al.*, 1990).

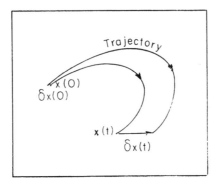

Fig. 8 Schematic diagram showing exponential divergence with changes in initial time

As an illustrative example, let us consider here two different types of time series generated by a purely chaotic model (logistic) of the form of $X(t + 1) = \mu X(t) (1 - x(t))$ and the logistic model plus non-linearly modulated sinusoids. This non-linear logistic model earlier proposed by May (1977) exhibits a remarkable range of dynamical behaviour depending on the values of μ. The results of Lyapunov exponents for both sets of data are displayed here. Figure 9 shows a positive Lyapunov exponent, confirming the existence of chaos.

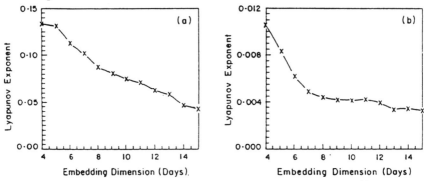

Fig. 9 Lyapunov exponent of theoretical generated model discussed in text

Wolf *et al.* (1985) developed an elegant and simple method of computing Lyapunov exponents from a one-dimensional time series. This algorithm, in particular, is quite robust for small and noisy data sets. The procedure for calculating the Lyapunov exponents are discussed in detail by Wolf *et al.* (1985), which follows the embedding dimension theory of Takens (1981), as discussed in the previous section. For a given point on a trajectory in n-dimensional phase space at time t the following one at $(t + 1)$ will be its image. A search process is conducted for the nearest neighbour of the point inside a sphere of radius R_{max}. After recording their mutual distance $X(t)$, the algorithm searches for the images of the two points. The ratio of the new distance $X(t + 1)$ to the original distance contributes to the sum in eqn. (6). The procedure is repeated starting from the point $t + 1$, until all the points of the trajectories are exhausted. As noted by Ott and Spano (1995) the nearest neighbour should lie insofar as possible along the same direction relative to the trajectory. Appropriate care should also be taken, such that the angle between the directions of the new nearest neighbour and the image of the previous one should not exceed a chosen maximum angle θ_{max} (Vassiliadis *et al.*, 1990).

Computation of the Lyapunov exponent requires an appropriate choice of control parameters embedding dimension, maximum and minimum scale parameters of attractor, the time lag to construct the phase space, and the time of trajectory evolution, i.e., the time the system spends in the attractor. In the present analysis, we have chosen an embedding dimension between 8 and 12, which is justified for a fractal dimension in the range of 4-6. For the delay time, we used that predicted by the autocorrelation function (in our case, 5 days) and for evolution time 14 to 42 days. These control parameters (scale parameters, evolution time and delay time lag (τ)) were appropriately varied to obtain successive stable estimates of λ. This exercise finally provides confidence that the convergence of the Lyapunov exponents is independent of run parameters. The Lyapunov exponent for a different number of iteration steps is shown in Fig. 10. The analysis indicates low positive values of the Lyapunov exponents over an approximately chosen range of values of control parameters. Although there are minor fluctuations in the Lyapunov exponent value, it is remarkably positive and stable. The Lyapunov exponent quantifies the nature of inherent irregularity (chaos) and rate of separation for an exponential divergence of LOD/AAM variations.

Kolmogorov entropy (K_2)

The Kolmogorov entropy (K_2) is a measure of the rate at which information about the state of the dynamical system is lost in the course of time, which provides the characteristic measure of chaotic motion. The concept of entropy essentially provides the distinction between random and deterministic systems. Greater entropy is often associated with more randomness and less with the system order. Entropy has also been shown to be a parameter that characterises chaotic behaviour. The quantification of entropy may therefore have two important implications. First, a finite entropy suggests that the data representing the underlying dynamics do

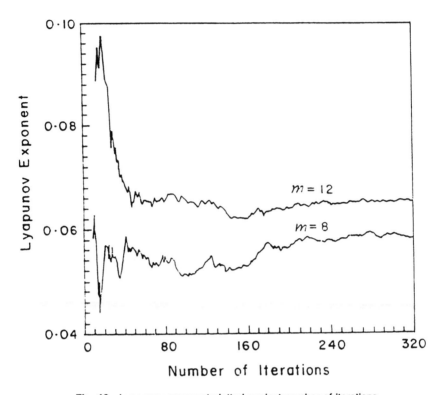

Fig. 10 Lyapunov exponent plotted against number of iterations

not represent a random system but rather a chaotic one and secondly, the system renders a characteristic time scale of the system dynamics which can be used as a time limit for its predictability.

For a dissipative physical system which is characterised by a low dimensional strange attractor (Tiwari *et al.,* 1992), an accurate forecast is possible only for a certain time interval.

Possible predictive time limit can be calculated using the dynamical quantity (K_2), which is an approximation to the Kolmogorov-Sinai invariant entropy (K_2). The Kolmogorov entropy is defined as (Vassiliadis *et al.,* 1990):

$$K_2 = \lim_{r \to 0} \frac{1}{\Delta t} \frac{C_m(r)}{C_{m+1}(r)} \qquad (7)$$

where Δt is sampling rate and $C_m(r)$ is the correlation function. In fact, K_2 entropy has the property that K_2 is zero for an ordered system and is infinite in a random system. K_2 is a finite constant > 0 for chaotic dynamics. It is noted that finite and convergent K_2 value and strange attractor dimension have been taken almost as proof of the presence of low dimension chaos in the dynamics of a system with irregular behaviour.

We have calculated the second order K_2 entropy following eqn. (7) and the method discussed in the previous section. Figure 11 shows the dependence of correlation entropy K_2 on the embedding dimension for original LOD data. Obviously, the value of K_2 for different values of distance r is seen to decrease with increase of embedding dimension and tends to be saturated for higher embedding dimensions. Obviously, Fig. 11 shows that the entropy is finite and positive ($K_2 > 0$), thus providing possible evidence of the chaoticity of LOD time series. The saturated value of K_2 for original LOD time series is 0.17-0.02 corresponding to a predictive time limit of about 7-8 days.

DISCUSSION

The coupled ocean-atmospheric dynamics can be modelled as low dimensional/order chaotic processes (Lorenz, 1963; Jin *et al.*, 1994; Chang *et al.*, 1994; Tziperman *et al.*, 1994). These workers have also suggested that non-linear interactions between seasonal cycles and interannual variations in coupled ocean atmospheric system can be a probable cause of irregularities/chaos in the El Nino/Southern Oscillations (ENSO). Recent studies have also demonstrated that the Earth's rotational dynamics can be viewed as a low dimensional chaotic dynamical system (Tiwari *et al.*, 1992, 1994, 1996). It is known that ocean atmospheric circulations and the Earth's rotational dynamics are intimately linked phenomena. The major source of non-linearity clearly arises from the response of the atmospheric to thermal forcing and from thermodynamic interactions between the atmosphere, ocean and the Earth's dynamical state. The broadband nature of LOD spectrum and clustering and splitting of spectral peaks apparently visible/submerged into background noise are some of the probable signatures/characteristics of the inherent non-linearity and chaos in the system dynamics (Tiwari *et al.*, 1994). The non-linear interaction of various frequencies taking place affects the LOD and atmospheric dynamics. In particular, it may be suggested that the non-linear resonances created by the seasonal cycle and inter- and intra-annual fluctuations due to thermodynamic property of the atmosphere-ocean system (possibly through a series of frequency locking and overlapping of the resonances) may eventually turn the system dynamics to a chaotic route. Understanding these physical interactions and a particular route of chaos might possibly enable causally related predictions to be made for the underlying dynamical system. This is one of the several possible conjectures regarding the predictability of coupled earth-ocean-atmospheric dynamics.

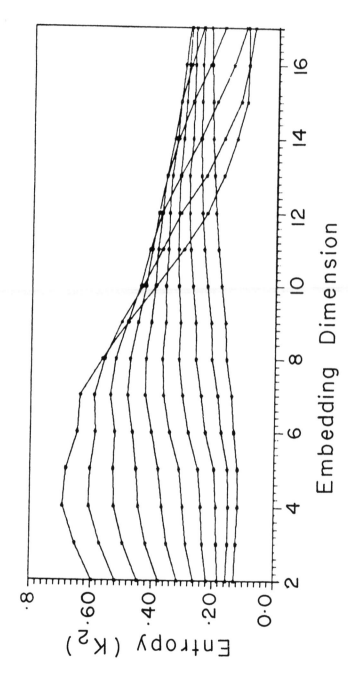

Fig. 11 K_2 Entropy of original LOD time series. Vertical scale shows variations of entropy (K_2) and horizontal scale embedding dimension

Evidence of chaos has been found in the dynamics of the solar system (Sussman and Wisdom, 1988), planetary motion (Laskar and Robutel, 1993) and in many other coupled earth-ocean-atmospheric systems (Lorenz, 1963). The present finding of deterministic chaos in the LOD/AAM system adds yet another evidence to this series of significant geophysical and astronomical studies.

CONCLUSIONS

In essence, several complexities still remain unresolved, but certain points emerge which provide a new reasoning: (i) dimensional analysis of the LOD/AAM attractor reveals a low dimensional attractor of the order of 5-6; (ii) LOD/AAM dynamics indicates signatures of a low positive Lyapunov exponent consistent with self-organised critical dynamical systems. The computed Lyapunov exponent is remarkably stable, indicating possible evidence of chaos in the LOD/AAM system, (iii) a probable source responsible for the exponential divergence (chaos) in LOD/AAM system could be the interaction of seasonal cycles with inter- or intra-annual variation due to thermodynamic properties of the atmosphere-ocean system; (iv) the presence of coherent time structures in chaotic LOD time series creates the physical premises for theoretically modelling coupled LOD system dynamics.

Ours is but one conjecture among several possible explanations of chaos in the LOD/AAM system. We hope this presentation may lead to stimulating analyses and discussion for further understanding of the complexity of the Earth's axial rotation and its linkages with the dynamics of component subsystems.

Acknowledgement: We thank Dr. V.P. Dimri, for appreciating this work and Dr. H.K. Gupta, Director, NGRI for kind permission to publish it. We also thank Shri K.N.N. Rao for help in computational work and Mr. V. Subrahmanyam for assistance in manuscript preparation.

REFERENCES

Abraham NB, Albono AM, Das B, Deguzman G, Gioggia RS, Pyccions GP, Tredicce TR and Yong S. 1986 Calculating the dimension of attractors from small data set. *Phys. Lett.,* **A114:** 217-221.

Baddii RG, Brossi B, De righettii, Ravani M, Gilberts S, Poliiti A and Reibig MA. 1988. Dimension increase in filtered chaotic signal. *Phys. Rev. Lett.,* **60:** 797-782.

Chang P, Bin Wang, Tim Li and Link Ti. 1994. Interactions between seasonal cycle and southern oscillation-frequency-entrainment and chaos in a coupled ocean atmospheric model. *Geophys. Res. Lett.,* **21:** 2817-2820.

Crowley and North GR. 1990. Abrupt climate change and extinction events in earth history. *Science*, **240**: 996-1002.

Delache P and Scherrer PH. 1983. Detection of solar gravity mode oscillations. *Nature*, **306**: 651-654.

Fuller WA. 1976. *Introduction to Statistical Time Series*. John Wiley and Sons, New York.

Grassberger P and Procaccia I. 1983a. Characterization of strange attractors. *Phys. Rev. Lett.*, **50**: 346-349.

Grassberger P and Procaccia I. 1983b. Estimation of the Kolmogorov entropy from a chaotic signal. *Phys. Rev.*, **A28**: 2591-2593.

Jin FF, Neelin JD, Ghil M. 1994. El Nino on the devil's staircase, annual subharmonics step to chaos. *Science*, **264**: 70-72.

Kaiser F. 1990. Non-linear dynamics and deterministic chaos: Their relevance for biological functions and behaviour. In: *Geoscience Relations. The Earth and its Microenvironment*. pp. 315-320. G.J.M. Tomassen *et al.* (eds.). Pydoc The Nethelands.

Laskar J and Robutel P. 1993. The chaotic obliquity of the planets. *Nature* **361**: 608-612.

Lorenz FN. 1963. Deterministic nonperiodic flow. *J. Atmos. Sci.*, **20**: 130-141.

Louterborn W and Porlitz U. 1988. Method of chaos physics and their application to acoustics. *J. Acoust. Soc. Am.*, **84**: 1975-1993.

May RM. 1976. Simple mathematical models with very complicated dynamics. *Nature*, **261**: 459-467.

North GR, Mengal JG and Short DA. 1983. Simple energy balance model resolving the seasons and the continents—Application to the astronomical theory of the ice ages. *J. Geophys. Res.*, **88(C11)**: 6576-6586.

Ott E and Spano M. 1995. Controlling chaos. *Physics Today*, May, pp. 34-40.

Roberts DA. 1991. Is there a strange attractor in magnetosphere? *J. Geophys. Res.*, **96**: 16031-16046.

Rosen RD and Salstein DA. 1983. Variation in atmospheric angular momentum on global and regional scales and the length of the day. *J. Geophys. Res.*, **88**: 5451-5470.

Rosen RD, Salstein DA and Nehrkorn T. 1991. Prediction of zonal wind and angular momentum by the NMC medium range forecast modelling during 1985-1989. *Mon. Weath. Rev.*, **119(1)**: 208-217.

Ruelle D. 1994. Where can one hope to profitably apply the idea of chaos ? *Physics Today*, July, pp. 24-30.

Saltzman B. 1979. A survey of statistical dynamical models of terrestrial climate. *Adv. Geophysics*, **20**: 183-304.

Sussman GJ and Wisdom J. 1988. Numerical evidence that the motion of Pluto is chaotic. *Science*, **241**: 433-437.

Takens F. 1981. Detecting strange attractors in turbulence. In: *Lecture Notes in Mathematics*, pp. 361-381, Rand, D.A. and Young, L.S. (eds.). Springer Verlag, Berlin.

Tiwari RK, Negi JG and Rao KNN. 1992. Attractor dimension in non-linear fluctuation of length of day time series. *Geophys. Res. Lett.*, **19**: 909-912.

Tiwari RK, Negi JG and Rao KNN. 1994. Strange attractor in non-linear fluctuations of length of the day (LOD) time series. Non-linear dynamics and predictability of geophysical phenomena. *Geophysical Monograph 83*, IUGG, vol. 18, pp. 61-67. W.I. Newman, A. Gabrielov and D.L. Turcotte (eds).

Tiwari RK, Negi JG and Rao KNN. 1996. Nature of chaos and limit of predictability of length of day (LOD) fluctuations. *J. Geophysics*, **18(2)**: 153-158.

Triantafylloy GN, Elsner JB, Lasearatos A, Kartitas C and Tsonis A. 1995. Structures and properties of the attractor of a marine dynamical system. *Math. Computer Modelling*, **21(6)**: 73-86.

Tsonis AA and Elsner JB. 1988. The weather attractor over very short time scales. *Nature*, **333**: 545-547.

Tziperman E, Stone L, Cance MA and Jarosh H. 1994. El Nino chaos, overlapping of resonances between the seasonal cycles and Pacific Ocean atmosphere oscillator. *Science* **264**: 72-74.

Vassiliadis DV, Sharma AS, Eastman TE and Papadopoulos K. 1990. Low dimensional chaos in magnetospheric activity from AE time series. *Geophys. Res Lett.*, **17**: 1841-1844.

Vio R, Cristiani S and Provenzale A. 1992. Time series analysis in astronomy. An application to Ovasar variability studies. *The Astrophysical Journal*, **301**: 518-530.

Wolf A, Swift JB, Swinney HL and Vastano JA. 1985. Determining Lyapunov exponents from a time series. *Physica*, **16D**: 285-317.

APPLICATION OF CATASTROPHE THEORY TO SOME NON-LINEAR GEOPHYSICAL PROBLEMS

R.K. Tiwari*

INTRODUCTION

Climate dynamics (glacial and interglacial) coupled with ocean and atmospheric processes, phase changes in mantle dynamics and frequent occurrence of earthquakes phenomena exhibit a catastrophic 'jump'. Linear and non-linear numerical and theoretical modelling of these critical processes have been extensively carried out for some years. However, difficulties have been encountered: First for instance, developing analytical models and solving them analytically does not provide expected results. Numerical-based techniques have been developed to overcome this problem. However, even after detailed numerical calculation over a range of parametric values, it is not possible to make general statements about the effect of varying parameters. Consequently, presentation of results in a coherent way becomes difficult. This is due to the fact that characteristic jumps in the amplitude of the system dynamics occur due to changes in combination of controlling parameters and not just due to change in a single parameter.

* Theoretical Geophysics Group, National Geophysical Research Institute, Hyderabad 500 007, India

Most of the geophysical/geological processes pertaining to the geo-bio-atmosphere-ocean system are highly complex and non-linear in nature. Developing and/or implementing new techniques to understand the nature of these physical processes and their interacting physical mechanism with respect to probable linkages to exogenic endogenic forcings, therefore, become essential for making prediction or at least understanding the qualitative nature of their sensitive critical physical behaviour. The various kinds of basic data (seismological, climatic, geochemical, geomagnetic etc.) analysis and their appropriate interpretation, combined with proper understanding of theoretical models, would render significant insight and shed light on the evolution of these processes. The catastrophe theory is ideally suited for studying such catastrophic phenomena. In the following sections, we discuss application of the catastrophe theory to study the response of two climate models, as a means for analysing the dynamical climate variability mechanism.

SOME FEATURES OF THE CUSP CATASTROPHE THEORY

The catastrophe theory is concerned with creation and annihilation of critical points (sudden discontinuities in the behaviour of a physical system) in response to small perturbations of the controlling parameters (Fig. 1). A potential function of a prototype model to be discussed in connection with the climate dynamics and earthquake may be written as (Poston and Stewart, 1978):

$$u_{\alpha\beta} = \frac{x^4}{4} - \alpha\,\frac{x^2}{2} - \beta x \qquad (1)$$

where x is a scalar valued function.

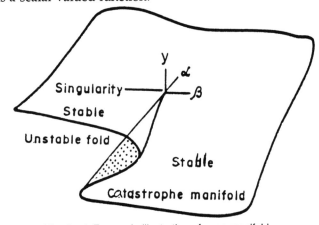

Fig. 1   A Zeeman's illustration of cusp manifold

For fixed α and β, the critical points are given by

$$\frac{\delta u_{\alpha\beta}}{\delta x} = \frac{\delta^2 u_{\alpha\beta}}{\delta x^2} = 0 \tag{2}$$

Equation (2) will have a minimum of one root and at most two real roots, depending on the values of α and β. Sudden changes in equilibria may implicitly be defined by eqn. (2), Namely

$$x^3 - \alpha x - \beta = 0 \tag{3}$$

$$3x^2 - \alpha = 0$$

Eliminating x from these equations gives the discriminant which represents a cusp-like curve and is called 'cusp catastrophe'.

$$\beta^2 = -\frac{4\alpha^3}{27} \tag{4}$$

Understanding the apparent unpredictability of the onset and termination of catastrophic climate phenomena (glacial/interglacial) is possible by studying the 'critical point' behaviour of the system dynamics. The glacial onset and terminations can be assumed as mathematical catastrophe. The state parameters x in the equation define the dynamic state of the system and manifest discontinuities in the system behaviour identifiable on the catastrophe surface. In particular, parameter α is related to the amount of energy released during the onset of climate catastrophe and parameter β is crucial for triggering the physical system. Bifurcation set of the cusp catastrophe is obtained by the cusp equation $4\alpha^3 + 27\beta^2 = 0$. Abrupt changes in any physical system described by the cusp catastrophe can be determined by this equation. A striking situation in climate dynamics has been noticed wherein quite a small insolation change due to eccentricity variations of the earth's orbit leads to large spectral peaks in spectrum of climate indicator time series. Several suggestions have been put forward to explain the mystery. The catastrophe model may identify limited number of important structurally stable ways in which smooth changes in the external forcing can give rise to large changes in the state parameters.

ANALOGY BETWEEN DYNAMICS OF ZEEMAN CATASTROPHE MACHINE AND VARIABILITY OF POTENTIAL ENERGY IN CLIMATE

To explain the variability in the energy state of climate evolution, one has to calculate the 'threshold' value of controlling parameter as defined by the bifurcation set of the catastrophe. But before that we draw attention to an interesting analogy between dynamics of Zeeman's catastrophe machine and climate dynamics which can be quite tempting to explain such behaviour. Climatic changes are triggered by internal as well as external feedback mechanisms. Absorbed heat energy by the earth from solar radiation is the source of external energy for the dynamical climate system (North *et al.*, 1990). It is the product of the amount of energy incident on the top of the atmosphere and absorptivity (albedo) of the earth's atmospheric column (Crowley and North, 1990). The physical system reacts in such a way as to minimise the potential energy within the constraints imposed by external conditions. Figure 2a shows the initial state of the system. The dynamical system begins with stored energy and a bifurcation takes place when an inflection point is created (Fig. 2b). New stable configuration in the form of local potential energy minimum emerges and eventually becomes deeper to the level of 'criticality' in the presence of feedback inputs. At this state the potential energy becomes so large that the first minimum coalesces with the central maximum. The equilibrium breaks down with an associated explosive release of energy. The process forms a point of inflection and causes the system to 'jump' to a new stable state (Fig. 2c to e). The climate machine flip suddenly to the new minimum (Fig. 2f). The whole sequence of coalescence of minima and maxima of the potential energy curve is a mathematical catastrophe, as demonstrated in Fig. 2a-g (Zeeman, 1976).

Fig. 2   Illustration of climate dynamics in terms of Zeeman's catastrophe machine

SIMPLE MATHEMATICAL CATASTROPHE MODEL OF PALEOCLIMATIC SEA-ICE-OCEAN-TEMPERATURE (SIOT) OSCILLATOR (MODEL I)

Non-linear differential equations forming dynamical systems can induce/ govern such changes in the presence of slow external forcing or stochastic forcing. In an interesting pioneering attempt Saltzman *et al.* (1981) proposed the SIOT oscillator model to describe a possible interaction between ice coverage and the ocean state as a coupled system. The main characteristics of the interactions between different system components of the SIOT model can be described in the following mathematical form (Saltzman *et al.*, 1981 and Nicolis, 1984).

$$d\theta/dt = -\alpha n + \beta\theta - \theta n^2 \tag{5}$$

$$dn/dt = -n + \theta \tag{6}$$

Here α and β are positive variables. Equation (5) represents the ocean surface heat balance and eqn. (6) stands for the mass balance of sea ice. In particular, the term describes the negative insulating effect of the sea ice on CO_2 and the positive feedback of CO_2 on itself due to atmospheric carbon dioxide (Nicolis, 1984). The non-linear term accounts for the non-linear restoring mechanism, effective only when the system is displaced far from the initial state. For a quantitative study of the complexities in the climate dynamics both the initial variables and parameters of the simple lumped oscillator in eqns. (5) and (6) are suitably scaled to non-dimensional quantities. These highly reduced dynamical systems ('toy model' in field vernacular) yielded an important set of ideas about how external forcing variability and internal dynamics might interact in controlling the climate variability spectrum. Indeed, at present, numerous other versatile physical models exist that account for abrupt changes as well as periods, quasi-period and periodic variations (Saltzman, 1979). Also, Matteucci (1989) based on the energy balance model has discussed these variations. Nevertheless, we feel that in order to derive some insight regarding the non-linear interactions in coupled systems, it might be interesting and worthwhile to discuss some of the properties of SIOT models in terms of the catastrophe theory.

To understand the sudden loss of stability and bimodal climatic behaviour, we recast eqns. (5) and (6) in terms of the catastrophe theory. This transfomation naturally follows from a more general theory of the non-linear dynamical system. Using the equilibrium condition in eqns. (5) and (6) and making necessary substitutions, the following cubic equation results:

$$\alpha^1 n - n^3 = 0 \tag{7}$$

where $\alpha^1 = (\beta - \alpha)$

This is evidently an incomplete cusp which arises in the absence of an external forcing. The unfolding of eqn. (3) is in the form of a non-transverse and structurally unstable cusp (Poston and Stewart, 1978). The solution of the eqn. (3) is given by the line $n = 0$ and the parabola $\alpha^1 - n^2 = 0$. This defines the fold catastrophe, the simplest of all forms (Fig. 3). The corresponding potential function will be given by:

$$\frac{n^4}{4} + \frac{\alpha^1 n^2}{2} = 0 \tag{8}$$

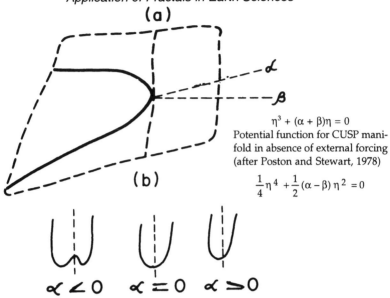

$$\eta^3 + (\alpha + \beta)\eta = 0$$

Potential function for CUSP manifold in absence of external forcing (after Poston and Stewart, 1978)

$$\frac{1}{4}\eta^4 + \frac{1}{2}(\alpha - \beta)\eta^2 = 0$$

Fig. 3 (a,b)—Illustration of incomplete and structurally unstable cusp manifold in the absence of external forcing terms and potential functions

Mathematically, the resultant behaviour is universal for a typical family of a functions of single parameter passing through a non-Morse singularity (Poston and Stewart, 1978). However, as an unfolding among even functions of those with $f(n) = f(-n)$, it depicts a unique stable single parameter with local family around $n = 0$ up to the usual diffeomorphism (Poston and Stewart, 1978).

Incorporation of the essential term r in the family of eqn. (3) leads to a full cusp catastrophe model as described by:

$$r + \alpha^1 - n^3 = 0 \tag{9}$$

where $r = (Q/Q_0)$ is normalised solar constant.

Here r is the 'external forcing' factor, which is related to the triggering of the physical system and α^1 is related to the amount of energy released during the onset of catastrophe. The magnitude of 'jump' is related to the state parameter. Figure 4 depicts the solution curve of the cusp model for a typical and scaled value of control parameters (Nicolis, 1984). Further, it may be noted that the variations of sea-ice extent with respect to the parameters α^1 and r are functionally related with each other. To understand the functional relationship between the state parameters and the control variables, let us suppose that the initial state of the climate was ice free (right side of Fig. 5). Lowering the solar constant moves the equilibrium solution along the point towards the dotted curve. Further decrease in the solar constant by small increments suddenly forces the equilibrium point to jump to the upper solution.

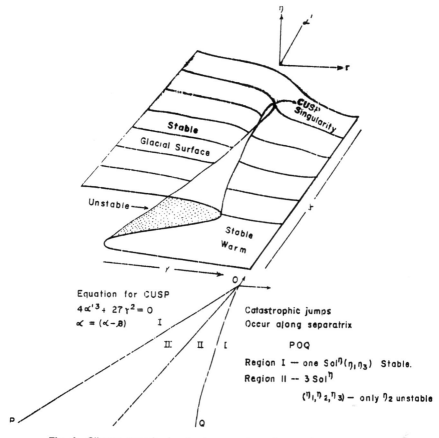

Fig. 4 Climate cusp for feedback parameter values and external forcing

Equation (9) has three solutions. The three real roots are generally arranged in descending order and in this sequnce the middle one is the unstable component while the other two roots are the stable ones. The unstable component of the system behaviour is sandwiched between the two stable states for which no physical explanation can be given. Figure 5 shows the mode response curve. It enables us to comment on a physical interpretation of the critical point. One can visualise that as r (solar radiation) decreases, the state parameters (the area of ice extent) increases and vice versa. For a critical value (of somewhere around –0.5) the model attains 'criticality' condition.

Thus we may conclude that analysis of existing non-linear SIOT oscillator model in terms of the mathematical catastrophe theory leads to a non-transverse, incomplete and structuraly unstable 'cusp'. Incorporating for the effect of the external perturbing term in the SIOT model, the new model presented here converges to a structurally stable cusp catas-

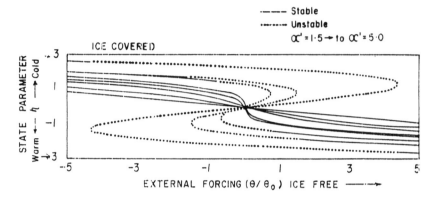

Fig. 5   Stereographic projection of three-dimensional surface on to two-dimensional map (illustration of full cusp using stacking procedure)

trophe and explains some natural features of ocean and terrestrial climate swings. The novel strategy uses unfolding of the fundamental cusp sigularity and also assumes apparent stochastic behaviour of the glacial onset and abrupt climate transitions. Analysis indicates that an increase of about 60-70 per cent in CO_2 level from its initial value leads to catastrophic jump when two coupled control parameters are critically and functionally related. A small variation in external forcing near criticality results in large climate transitions. The model behaviour is in accordance with the qualitative nature of onset of long-term glaciation and recent termination of Younger Dryas.

FORCED HARMONIC CLIMATE OSCILLATORS (MODEL II)

One of the central questions in climate dynamics is concerned with the relative role of internal and/or external climate drivers. Many climatic episodes, including Quaternary glaciations, present a cyclic character. The characteristic time is strongly correlated with external forcing, as are the Earth's orbital variations. However, the balance equations of the prin-ciple variables predict that the coupling of such external forcing (solar insolation) due to the Earth's eccentricity to climate dynamics is too small to produce an observable peak around 100 kilo years (kyr) in the climate spectra. Hence it is difficult to understand how such small amplitude insolation disturbances due to the Earth's eccentricity could cause anoma-lous response in the form of major climate change. To unravel this mystery, an interesting and appropriate representation of such coupled climate models might possibly be represented by the following Duffing equation:

$$\frac{d^2 n}{dt^2} + \omega_0 n^2 = q \cos \omega t + n\gamma - \beta - \frac{dn}{dt} - \alpha n^3 \tag{10}$$

where ω_0 is natural frequency of internal climate dynamics and ω is forcing frequency (in this case eccentricity cycle) of the system, q represents the amplitude of the forcing term and α, β and γ are the coefficients of non-linearity, damping and source term respectively. Equation (10) differs from the earlier physical climate model of Moritz (1979) in mainly three respects. Here, eqn. (10) includes additional terms, e.g. (i) natural frequency ω_0 of the system, (ii) coupling with external frequency term ω and (iii) higher order non-linearity. In fact, eqn. (10) presumably represents coupled climate system dynamics. It provides more comprehensive representation of temporal climate variations. The comprehension is justified since external forcing (orbital variations) plays a crucial role in modulating climate dynamics. The effect of amplitude and external frequencies, e.g. $q \cos \omega t$ should therefore be studied in combination with other parameters. The Duffing oscillator explains some of the real features of climate dynamics.

The solution and properties of the Duffing oscillator have been well studied in connection with many general physical processes. The solution of the Duffing eqn. (10) leads to the form of cusp catastrophe (Holmes and Rand, 1976) given by

$$y^3 + Py + Q = 0 \tag{11}$$

where P and Q are the control parameters.

We can derive climate control parameters P and Q for a climate model following Guckenheimer and Holmes (1983) and study the response function, as

$$P = -3g^2 + (M_3/M_1)$$

$$Q = -2g^3 + (M_3 g/M_4)$$

where $\quad g = -(M_3/3M_1)\, 8\Omega/9\alpha$

$$M_1 = 9/16\alpha^2 , M_2 = (3/2)\Omega\alpha$$

$$M_3 = \Omega^2 + \omega^2\beta , M_4 = -q^2, \Omega = \omega^2_0 - \omega^2 \qquad \text{for } \varepsilon = 1.$$

Here phase can be given by $\tan \theta = QWS/(4\Omega + 3\alpha\gamma^2)$

Using the probable climate coefficients in eqn. (11), we can study the effect of external forcing (eccentricity cycle variation) with changes of its amplitude with other parameters. The following coefficient values are chosen initially:

$$\alpha = af g_1 ; \beta = b\delta ; \gamma = aef.$$

$$a = 4.20 \times 10^{-9} \text{ K}^{-1} \text{ S}^{-1}$$

$$b = 6.2720 \times 10^{-8} \text{ S}^{-1}$$

$$c = 1.10 \times 10^{-6} \text{ S}^{-1}$$

$$e = 4.14 \times 10^{-9} \text{ K}^{-1} \text{ S}^{-1}$$

$$f = 4.68 \times 10^{-9} \text{ K}^{-1} \text{ S}^{-1}$$

$$g = 10^{-4} \ldots 10^{-2}$$

Figure 6 depicts the variations of amplitude of the state parameter (y) of system eqn. (11) with respect to time for different values of forcing amplitude term q. From Fig. 6, it may be inferred that even a small change in forcing amplitude q produces an appreciable change in system response. The system attains maximum amplitude at around 100 ky. For a slight change in q, the forcing amplitude system reaches a critical value, some 'threshold' value, leading to catastrophic jumps. Here the model also indicates typically three branches of solutions. The solid and broken lines represent respectively, the stable and unstable conditions. The upper and lowermost branches (solid line) exhibit minima corresponding to the stable conditions; they correspond to warm climate and a completely ice-covered earth respectively. These two stable branches are separated by an unstable one (broken line which corresponds to maxima). This, perhaps, represents the chaotic transitional phase.

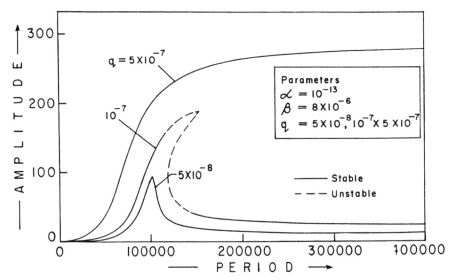

Fig. 6 Variations of system amplitude with external forcing parameter q (here earth's eccentricity cycle). Solid lines show stable solution and asterisked lines unstable solution

Figure (7) shows the variation of the system amplitude for various values of the non-linearity term parameter. It would be seen from Fig. 6, that there is also appreciable change in the amplitude of system response at frequency $\omega = \omega_0$. Apparently non-linearity leads to the 'criticality' at the resonant frequency, i.e., when $\omega = \omega_0$. It may therefore be inferred that when internal and external forcings are equal, a small variation in external forcing combined with non-linearity via a 'link mechanism' plays a crucial role in triggering the climate catastrophe.

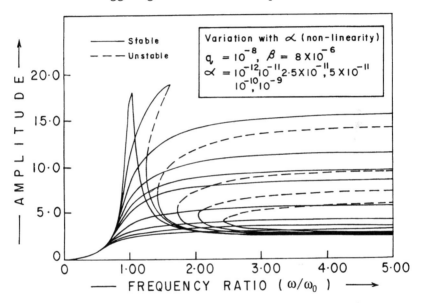

Fig. 7 Variations of system amplitude with non-linearity parameter. Solid lines show stable solution and asterisked lines unstable solution

CONCLUSION

The salient features of the present study can be summarised as follows: (i) The evolution of climate can be described by a cusp catastrophe whereby the stable and unstable surfaces can be classified by the singularities into catastrophe boundaries. (ii) This indicates a possible 100 kyr internal oscillation in climate dynamics coupled with external eccentricity forcing cycle. Even small variation in the amplitude of external forcing amplitude (small insolation variation due to eccentricity of the earth's orbit) may dramatically amplify the amplitude of the system dynamics as observed in geological records. (iii) Catastrophe model provides a viable alternative perspective to study and resolve some features of climate variability. In fact, behaviour of the cusp catastrophe model is very close to the generally accepted natural climate behaviour and can be applied without invoking a new and different mechanism.

Future effort should take into account the more precise values of coefficients to test the viability of this model which ultimately is capable of predictions matching with observations.

ACKNOWLEDGEMENT

We thank Dr. V.P. Dimri, for appreciation of this work and Dr. H.K. Gupta, Director, NGRI, for kind permission to publish it. We also thank Shri K.N.N. Rao for help in computational work and Mr. V. Subrahmanyam for assistance in manuscript preparation.

REFERENCES

Guckenheimer J and Holmes P. 1983. Non-linear oscillations, dynamical systems and bifurcations of vector fields. *Applied Mathmetical Science* 42, Springer-Verlag, Berlin.

Holmes PJ and Rand DA. 1976. Bifurcations of the Duffing equation: An oscillation of catastrophe theory. *J. Sound Vibration,* **44(2):** 237-253.

Matteucci G. 1989. Orbital forcing in a stochastic resonance model of the Late Pleistocene climate variations. *Climate Dynamics,* **3:** 179-190.

Moritz RE. 1979. Non-linear analysis of a simple Sea-Ice-Ocean temperature oscillator model. *J. Geophys. Res.,* **84(68):** 4916-4920.

Nicolis C. 1984. A plausible model for the synchroneity or the phase shift between climate transitions. *Geophys. Res. Lett.,* **11(6):** 587-590.

North *et al.* 1983. Simple energy balance model resolving the seasons and continents: Application to the astronomical theory of ice ages. *J. Geophys. Res.,* **80:** 6576-6586.

Poston T and Stewart Ian 1978. *Catastrophe Theory and Its Applications.* Pitman, London.

Saltzman B, Sutera A and Evenson A. 1981. A stochastic stability of a simple autoscillatory climate feedback system. *Atmosph. Science,* **38:** 494-503.

Zeeman EC. 1976. Catastrophe theory. *Scientific American,* **234:** 65-83.

APPLICATION OF FRACTAL DIMEN-SION IN STUDYING GEOMORPHIC PROCESSES—A CASE-STUDY FROM HISTORICAL CLIMATE DATA SET

Dhananjay A. Sant*

INTRODUCTION

A dynamic geomorphic system is characterised by (a) the state of the geomorphic system and (b) the processes acting upon the system. Various sets of landforms together determine the state of the geomorphic system. The erosional surfaces (regional landforms) that occur at various elevations are further characterised by minor landforms that in turn determine the surface texture (topography) of the erosional surface.

Two important processes that influence the state of the geomorphic system are: (1) the climate, and (2) the tectonics. They are also called 'State-Changing Processes'. The state-changing processes act at different rates, whereby the system is forced to cross an external threshold. State changing processes independently or in combination modify the landscape over a period of time. To quantify and differentiate the role of tectonics and climate is a present-day challenge to the geomorphologist.

* Department of Geology, Faculty of Science, M.S. University of Baroda, Vadodara 390 002, India

The present paper is restricted to a single process—the climate. Emphasis is given to understanding the dynamics of climate during historical time. A classic monsoon region (the Indian subcontinent) was selected in which the wet/dry seasonal pattern is caused by planetary atmospheric circulation features. Analysis focusses on obtaining a fractal dimension for three major components of the climate—temperature, pressure and precipitation. The climatic uncertainties are quantified by proposing a climate predictability index. The predictability indices are observed to change as the climate dynamics change from one season to the next. Further, since the predictability index gives a single dimensionless number for each process, it can be used to roughly quantify the interplay between temperature, pressure and precipitation.

The shift in emphasis from fractal dimension to predictability index may itself be useful since the latter concept is more initiative. Instead of thinking in terms of fractal dimension and associating it with implications for the time series, it is much more straightforward to go directly to the concept of predictability indices in the context of climate. The basic concepts would have applicability in other areas also.

CLIMATE PREDICTABILITY INDEX

Fractal dimension analysis of a geophysical time series has a long history (Hurst *et al.*, 1965; Mandelbrot and Wallis, 1969; Flugeman and Snow, 1989; Hsui *et al.*, 1993; Turcotte, 1992). However, these analyses have been mainly restricted to obtaining the fractal dimension of various time series. This paper attempts to go one step further and link these dimensions to the dynamics of the time series. Climate over any continent largely comprises four major components—geographic parameter, temperature, pressure and precipitation. Geographic parameters (latitude, longitude, and distance from sea and height) have not changed significantly and therefore remain constant. The time series for three climatic variables (temperature, pressure and precipitation) have already been shown to correspond to a Brownian motion (Mandelbrot, 1983).

Consider a discrete time series given by $x(t_i)$ (i = 1, 2, ...N). Here t denotes the time and x the amplitude of the variable under consideration (temperature, pressure or precipitation in our case). For a fractional Brownian motion, the amplitude increments $x(t_j) - x(t_i)$ have a Gaussian distribution with variance (Voss, 1985):

$$<[x(t_j) - x(t_i)]^2 > ~ (t_j - t_i)^{2H} \tag{1}$$

where the symbols $< >$ denote the average over many samples of $x(t)$. The parameter H is called the Hurst exponent and takes values between

0 and 1. If $H = 0.5$, we obtain the usual Brownian motion. The Hurst exponent is related to the fractal dimension D of the time series curve by the formula given as

$$D = 2 - H \qquad (2)$$

If the fractal dimension D for the time series is 1.5, we again get the usual Brownian motion. In this case, there is no correlation between amplitude changes corresponding to two successive time intervals. Therefore, no trend in amplitude can be discerned from the time series and hence the process is unpredictable. However, as the fractal dimension decreases to 1, the process becomes more and more predictable as it exhibits 'persistence', that is, the future or past trend is more and more likely to follow an established trend (Turcotte, 1992). As the fractal dimension increases from 1.5 to 2, the process exhibits 'antipersistence', that is, a decrease in the amplitude of the process is more likely to lead to an increase in the future or past. Hence, the predictability again increases. This scale independent unit thus gives us predictability of the acting process.

Fractal dimensions of the time series for three climatic variables—temperature (D_T), pressure (D_P) and precipitation (D_R)—are obtained by using rescaled range (R/S) analysis (Mandelbrot and Wallis, 1969), a conventional technique used for geophysical time records. Any other method would be equally adequate. The predictability indices for temperature, pressure and precipitation are:

$$PI_T = 2\,|\,D_T - 1.5\,|, \quad PI_P = 2\,|\,D_P - 1.5\,|, \quad PI_R = 2\,|\,D_R - 1.5\,| \qquad (3)$$

Here $|D|$ denotes the absolute value of the number D. The climate predictability index (PI_C) is thus defined by a set of predictability indices of temperature (PI_T), pressure (PI_P) and precipitation (PI_R) as

$$(PI_C) = (PI_T, PI_P, PI_R) \qquad (4)$$

If one of these indices is close to zero, then the corresponding process approximates the usual Brownian motion and is therefore unpredictable. If it is close to one, the climate is very predictable. The possibility to somehow combine these three indices into a single number using an appropriate norm was raised, but this may not be appropriate if even one of the processes is quite independent of the others. One important factor, which has not been explicitly included in making up the PI_C, is the geographic parameter of the region. However, in many cases it is possible that one of the aforesaid indices already includes the effect of the geographic parameter implicitly.

APPLICATIONS

The Indian subcontinent lies at the heart of the classic monsoon region and is the area most sensitive to monsoon fluctuation. Twenty-five measuring stations located throughout India were studied. The temperature, pressure and precipitation time series for these stations were obtained from the Global Historical Climatology Network (GHCN) data set (Vose *et al.*, 1985). In India low summer heat over the Tibetan plateau and northern India causes the strongest monsoon inflow, the South-West monsoon, during June to September (Fig. 1). While in winter the intense high pressure caused by the cooling of Siberia causes outflow of air that is partially blocked by the Himalayan ranges (Wasson, 1995; Lal *et al.*, 1994). As a result the Indian subcontinent is subjected to the North-East monsoon during the months of October to November (Fig. 2). We first calculated mean temperature, mean pressure and mean precipitation time series separately from the GHCN data for the two aforesaid periods. Fractal dimensions for these time series were then calculated using R/S analysis. From these fractal dimensions, the PI_C was calculated for all stations—one for the June-September period and the other for October-November period.

Fig. 1 Monsoon dynamics during June-September

Fig. 2 Monsoon dynamics during October-November

We were particularly interested to see the behaviour of PI_R with respect to seasons. It was found that PI_R changed with season in most of the stations. However, to overcome the possible errors involved in the calculation of PI_C from the time series, we restricted our attention to those stations where there was a significant change (greater than or equal to 0.4) in the PI_R. Four stations (Madras, Veraval, Agra, Nagpur) exhibited such a behaviour (Table 1), where precipitation is fairly predictable during the June-September period and becomes totally unpredictable during the October-November period. Thus, if the corresponding fractal dimensions and predictability indices are different, as in this case, then it is most likely that the region is influenced by more than one climatic dynamics. Consequently, in such cases, calculations done using mean yearly data are suspect. The equation of dynamics of all the four stations can be formulated as follows. PI_R of these four stations is strongly influenced by temperature, pressure and geographic parameter.

In the case of Madras and Veraval (both coastal stations) there is no change in PI_P. Hence it influences PI_R in a similar fashion in both periods. However, there is a significant change in PI_T. This is what causes the change in PI_R (Fig. 3).

CONSTANT

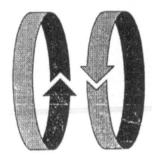

$$PI_T$$

$$PI_R$$

Fig. 3   Rainfall dynamics for Madras and Veraval (coastal stations) during South-West and North-East monsoon

Table 1   Values of predictability index paramaters at four stations

Station	Location	Jun. to Sep. PI_C (PI_T, PI_P, PI_R)	Oct. to Nov. PI_C (PI_T, PI_P, PI_R)
Madras	13.00° N 80.18° E	(0.6, 0.8, 0.4)	(0.1, 0.7, 0.0)
Veraval	20.90° N 73.37° E	(0.9, 0.4, 0.6)	(0.4, 0.4, 0.1)
Agra	27.02° N 78.00° E	(0.3, 0.7, 0.5)	(0.3, 0.2, 0.0)
Nagpur	21.10° N 79.05° E	(0.2, 0.6, 0.4)	0.2, 0.2, 0.0)

In the case of Agra and Nagpur (in central India), the situation is different. Here, there is no change in PI_T with the season. However, there is significant change in PI_P, which affects PI_R (Fig. 4).

Since the SW-monsoon is most important for India, we take stations for which precipitation during the monsoon is unpredictable (PI_R less than or equal to 0.1). There are four such stations (Bikaner, Jodhpur, Ahemdabad and Pune : Table 2).

For these stations both PI_T and PI_P are on the whole predictable. However, three of the stations (Bikaner, Jodhpur and Ahmedabad) are located in NW India, where the sinking limit of the SW-monsoon cell is located, resulting in unpredictability in precipitation (Das, 1962). The fourth station

CONSTANT

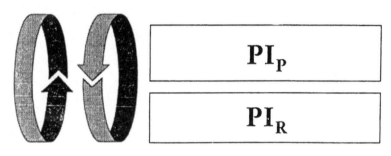

Fig. 4 Rainfall dynamics for Agra and Nagpur (inland stations) during South-West and North-East monsoon

Table 2 Predictability indices at four locations

Station	Location	Jun. to Sep. PI_C (PI_T, PI_P, PI_R)
Bikaner	28.00° N 73.30° E	(0.6, 0.4, 0.1)
Jodhpur	26.00° N 73.02°E	(0.2, 0.5, 0.0)
Ahmedabad	23.07° N 72.63° E	(0.8, 0.8, 0.0)
Pune	18.53° N 73.85° E	(0.4, 0.5, 0.0)

(Pune) is in the rain-shadow region of the SW-monsoon (Mishra and Rajaguru, 1995), which again accounts for its low PI_R. Thus the geographic parameter strongly affects their PI_R (Fig. 5).

DISCUSSION

The cases discussed above suggest that the PI_C captures some of the gross features of the climatic dynamics of the station. It further helps us in understanding how the various components of the climate interact with one another. The PI_C can be most useful in developing climatic models for the region. In case of climate predicting models, one should

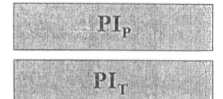

CONSTANT

BIKANER,
AHMEDABAD & PUNE
Geographic Parameter

$$PI_R$$

Fig. 5 Rainfall dynamics for Ahmedabad, Bikaner and Pune during the South-West monsoon

avoid stations with low PI_C, as data from such station would have random amplitude variations that are most probably caused by local conditions specific to that station. Studies also highlight that regional climatic models do not predict local events. Such anomalous stations can skew the entire model.

In the case of palaeoclimatic models interpreted from detailed field and laboratory studies of deposited sediments, many times two nearby sections give a different interpretation. Also, the lithosection represents large characters of the local event (flood/drought) that deposits sediments rather than overall climate. To quantify the high resolution data, their dynamics and to differentiate local and regional characters is the present-day task. In such cases PI_C would be very useful to understand the dynamics and feedback systems of global climate. Large erosional surfaces over the continent could be used as proxies for local and regional climate. Quantifying the same would give a better understanding.

We also believe that the shift in emphasis from fractal dimension to predictability index may itself be useful since the latter concept is more intuitive. Instead of thinking in terms of fractal dimension and then making the association with its implications for the time series, it is much more straightforward to directly go to the concept of predictability. Finally, even though we have used the predictability indices in the context of climate, the basic concept has applicability in other fields also.

REFERENCES

Das PK. 1962. *Tellus* **14**: 212-220.

Flugeman RH (Jr.) and Snow RS. 1989. *Pure Appl. Geophys.* **131**: 307-313.

Hsui AT, Rust KA and Klein GD. 1993. *J. Geophys Res.,* **98B**: 21963-21967.

Hurst HE, Black, RP and Simaika, YM. 1965. *Long-Term Storage: An Experimental Study.* Constable, London.

Lal M, Cubasch V and Santer S B. 1994. *Curr. Sci.* **66(6)**: 430-438.

Mandelbrot BB. 1983. *The Fractal Geometry of Nature.* WH. Freeman, New York.

Mandelbrot BB and Wallis JR. 1969. *Water Resources Res.,* **5**: 321-340.

Mishra S and Rajaguru SN. 1995. In: *Paleomonsoon from Desert Margins.* IGCP-349, Jaipur. Abstract, p.14.

Turcotte DL. 1992. *Fractals and Chaos in Geology and Geophysics.* Cambridge Univ. Press, NY.

Vose R S et al. 1985. *The Global Historical Climatology Network: Long-Term Monthly Temperature, Precipitation, Sea Level Pressure, and Station Pressure Data* (National Oceanic and Atmospheric Administration, Asheville).

Voss RF. 1985. In: *Scaling Phenomena in Disordered Systems.* Pynn, R and Skjeltrop, A. (eds.). Plenum, NY.

Wasson RJ. 1995. In: *Quaternary Environments and Geoarchaeology of India* (Rajaguru volume) Mem. 32 Geol. Soc. Ind.

Subject Index

Printed and bound by CPI Group (UK) Ltd, Croydon, CR0 4YY

23/10/2024

01777667-0010